D0630662

THE REVOLUTIONARY GENIUS OF PLANTS

The Revolutionary Genius of Plants

A New Understanding of Plant Intelligence and Behavior

Stefano Mancuso

ATRIA BOOKS

New York London Toronto Sydney New Delhi

An Imprint of Simon & Schuster, Inc.
1230 Avenue of the Americas
New York, NY 10020

English language translation by Vanessa Di Stefano
Originally published in Italy in 2017 by Giunti Editore S.p.A. as *PLANT REVOLUTION Le piante hanno già inventato il nostro futuro*

First Atria Books hardcover edition August 2018

Interior design by Kyoko Watanabe

Manufactured in China

10 9 8 7 6 5 4 3 2 1

Library of Congress Cataloging-in-Publication Data is available.

ISBN 978-1-5011-8785-8
ISBN 978-1-5011-8787-2 (ebook)

To Annina

Contents

Preface

It is my impression that most people don't really understand how important plants are for human existence. Of course, everyone knows—or at least I hope they do—that we are able to breathe because of the oxygen produced by plants and that the entire food chain, and thus the food that nourishes all animals on earth, relies on plants. But how many people realize that oil, coal, gas, and all the so-called nonrenewable energy resources are nothing more than another form of energy from the sun that was trapped by plants millions of years ago? Or that the active ingredients in many of our most important medicines come from plants? Or that wood, with all its amazing characteristics, is still the most widely used building material in much of the world? Our lives, as well as those of every other animal on this planet, depend upon the plant world.

You might think that we would already have discovered everything there is to know about organisms that are so important to the survival of mankind and on which a large part of our economy depends, but we're nowhere near that point. For example, in 2015 alone, 2,034 new plant species were discovered, and they were not microscopic plants that had somehow escaped the eyes of botanists. One of them, the *Gilbertiodendron maximum*, is a tree native to the Gabonese rain forest that is about 150 feet tall, with a trunk that can stretch to 5 feet in diameter, and that can weigh over 100 tons. And 2015 was not unusual; over the past decade, the number of new plant species discovered has exceeded 2,000 each year. Uses for more than 31,000 different species have been documented, including nearly 18,000 that have medicinal applications; 6,000 in our food supply; 11,000 in textile fibers and building materials; 1,300 for social purposes (including in religious rites and as recreational drugs); 1,600 as energy sources; 4,000 as food for animals; 8,000 for environmental purposes; and 2,500 as poisons. The sums soon add up: about a tenth of all plant species have an immediate use for humankind. Even if we recognize the central role plants play in our everyday lives—from the coffee in our cups to this book in your hands—the things that plants can teach us are less well understood. From materials to energy autonomy, from resistance capacity to adaptation strategies, from time immemorial plants have already provided the best solutions to most of the problems that afflict humanity.

Between four hundred million and one billion years ago, unlike animals, which had to move around to find their food, plants took an evolutionarily opposite direction: they remained in place, getting all the energy necessary to survive from the sun and adapting their bodies to resist predation and outmaneuver the other innumerable constraints arising from being rooted to the ground. That was no easy task. Think how difficult it would be to stay alive in a hostile environment without the ability to move. Imagine you are a plant, surrounded by insects, herbivorous animals, and predators of all kinds, unable to escape from them. The only way to survive would be to become virtually indestructible.

To counteract the problems associated with predation, plants have evolved in unique and unusual ways, developing solutions so different from those of animals that they have become the very epitome of diversity (indeed, plant species are so diverse that they might as well be aliens). Many of the survival solutions developed by plants are the exact opposite of those developed by the animal world. What is white for animals is black for plants and vice versa. Animals move, plants do not; animals are fast, plants are slow; animals consume, plants produce; animals make CO_2, plants use CO_2. But the most decisive contrast is also the least known: the difference between concentration and diffusion. Any function that in animals is concentrated in specialized organs is spread throughout the entire body of plants. This is a fundamental structural reason for why plants appear so different from us.

Our approach to engineering and design has typically been to replace, expand, or improve existing human functions. Humans have always attempted to replicate the basics of animal organization in the construction of their instruments. Take the computer, for example. It is based on ancestral schemata: a processor, which represents the brain and has the function of governing the hardware, plus hard drives, RAM for speeding access to memory, video, and sound cards. Everything that man designs tends to have, in a more or less overt way, a similar structure: a central brain that governs the organs that perform its commands. Even our societies are based on this same centralized, hierarchic, and archaic design, a model whose only advantage is to provide quick responses—not always correct ones—but that is very fragile and, as we shall see, not innovative or always effective.

Even though they have nothing akin to a central brain, plants exhibit unmistakable attributes of intelligence. They are able to perceive their surroundings with a greater sensitivity than animals do. They actively compete for the limited resources in the soil and atmosphere; they evaluate their circumstances with precision; they perform sophisticated cost-benefit analyses; and, finally, they define and then take appropriate adaptive actions in response to environmental stimuli.

Plants embody a model that is much more durable and innovative than that of animals; they are the living representation of how stability and flexibility can be combined. Their modular, diffused construction is the epitome of modernity: a cooperative, shared structure without any command centers, able to flawlessly resist repeated catastrophic events without losing functionality and adapt very quickly to huge environmental changes. The complex anatomic organization and key features of plants require a well-developed sensory system that enables the organism to efficiently explore the environment and react promptly to potentially damaging events. In order to exploit environmental resources, plants make use of, among other things, a refined root system made up of apexes able to detect and monitor concurrently and continuously at least fifteen different chemical and physical parameters in the soil. It is no coincidence that the Internet, the very symbol of modernity, is built like a root system. When it comes to robustness and innovation, nothing can compete with plants. Their way of functioning is a useful model, especially in an age when the ability to perceive change and find innovative solutions has become a requirement for success. We would do well to bear this in mind when planning for our future as a species.

Maple trees produce dry fruits called samaras that have a membranous wing useful for harnessing the power of the wind.

CHAPTER 1

MEMORIES WITHOUT A BRAIN

Memory: The mental faculty or power that enables one to retain and to recall, through unconscious associative processes, previously experienced sensations, impressions, ideas, concepts, and all information that has been consciously learned.

—***MOSBY'S MEDICAL DICTIONARY*****, eighth edition**

We have an immense memory, present in us without our knowledge.

—**DENIS DIDEROT**

PREVIOUS PAGE: We are used to identifying plants only by the parts that are visible coming out of the earth. In fact, the root system represents at least half of the body of a plant, and it is most likely the more interesting half.

ANIMALS OR PLANTS: EXPERIENCE TEACHES

After years spent investigating the many aspects of plant intelligence, I have been consistently surprised and fascinated by plants' clear capacity for memory. Maybe that sounds strange, but think about it for a moment. It isn't too difficult to imagine that intelligence is not the product of one single organ but that it is inherent to life, whether there is a brain or not. Plants, from this point of view, are the most obvious demonstration of how the vertebrate brain is an "accident," evolved only in a very small number of living beings—animals—while in the vast majority of life, represented by plant organisms, intelligence—the ability to learn, understand, and react successfully to new or trying situations—has developed without a dedicated organ.

It is impossible to learn without memory, and the ability to learn is one of the requirements of intelligence. Living things are generally capable of learning from experience, and plants are no exception to this principle; they respond in ever more appropriate ways when known problems recur throughout their existence. This could not happen without memory, the ability to store, somewhere in the organism, the relevant information needed to overcome those specific obstacles. But do not expect to hear others speak openly about memory when referring to the many plant activities that are similar to those that in animals require the use of the brain. When discussing plants, which do not have a brain, other terms are usually used: acclimatization, hardening, priming, conditioning, all of them linguistic acrobatics coined over the years by scientists to avoid the use of the comfortable and simple word *memory*. All plants are capable of learning from experience and therefore have memorization mechanisms. If you submit a plant, for example an olive tree, to a stress such as drought or salinity, it will respond by implementing the necessary modifications to its anatomy and metabolism to ensure its survival. Nothing unusual in that, right? If, after a certain amount of time, we submit the same plant to the exact same stimulus, perhaps with an even stronger intensity, we notice something that is surprising only on the surface: this time, the plant responds more effectively to the stress than it did the first time. It has learned its lesson. Somewhere it has preserved traces of the solutions found and, when there was a need, has quickly recalled them in order to react more efficiently and accurately. In other words, it learned and stored the best answers in its memory, thereby increasing its chances of survival.

PLANTS DO NOT HAVE A SHORT-TERM MEMORY

Many aspects of plant life that have important similarities to the animal world have a well-established, if not long, history of being studied (I am thinking of intelligence, communication skills, the ability to develop defense strategies, behavior, and so on). But in the case of memory, comparative tests started only in quite recent times. The first person to study it, however, was so significant that he was worth the long wait: we are talking about Lamarck. Or rather, Jean-Baptiste Pierre Antoine de Monet, Chevalier de Lamarck (1744–1829), because only his full name can truly suggest the importance of his activities as a scientist. Like other naturalists of his time, the father of biology—in the literal sense of the word, having coined the term itself—became interested in plant life, especially the phenomena of the rapid movements typical of the so-called sensitive plants (plants that respond to certain stimuli in an immediately obvious way). In particular, for much of his career he showed a keen interest in understanding the exact workings of the touch-sensitive mechanisms of the small leaves of the *Mimosa pudica*. It has to be said that even today we do not have a clear idea of how and why this plant's movements happen.

You may be familiar with the *Mimosa pudica*: today it is widely available in supermarkets. However, for those who have never seen it, it is an unusual and elegant plant that, as its name indicates, gently closes its leaves in an action of extreme modesty when subjected to some external stimulus (for example, if you touch it). Because of this immediate response, so rare in the plant world, this plant, which is native to the tropical regions of North, South, and Central America, aroused great interest when it arrived in Europe. Scientists such as

The *Mimosa pudica* in bloom. The pink inflorescence features many stretched stamens that give the flowers their feathery look.

Robert Hooke (1635–1703), the famous English microscopist who first saw and described a cell, and the French doctor René Joachim Henri Dutrochet de Néons (1776–1847), considered the father of cell biology, studied this plant. So for some years the *Mimosa pudica* was a true botanical star.

Its charm did not escape the Chevalier de Lamarck, who learned more about it through performing countless experiments and studying its behavior in situations that were, to say the least, original. But it was above all one detail that struck Lamarck: the fact that, when subjected to repeated stimulation of the same nature, there came a point when the leaves did not respond anymore and totally ignored each subsequent stimulation. Lamarck was right when he attributed this interruption to "fatigue." Basically, after repeated closures of the small leaves, the plant had no more energy available for further movements. Something similar to what happens in animal muscular activity, which cannot continue indefinitely and is limited by the amount of energy available, also happens in the *Mimosa pudica*. But not always.

Lamarck noted that sometimes, after frequent stimuli of the same nature, the "subject" stopped closing the leaves well before having exhausted its energy. That puzzled him; he could not understand the reason for this seemingly unpredictable behavior until one day he came across an ingenious experiment performed by the French botanist René Desfontaines (1750–1833), which seemed to answer his questions.

Desfontaines engineered an unusual experiment: he asked one of his students to take a large group of plants for a long tour of Paris by carriage and to scrupulously monitor their behavior. Above all, he had to pay attention to when they closed their leaves. The student, whose name we do not know and who was obviously accustomed to the extravagant demands of his teacher, did not bat an eye. He arranged several jars of *Mimosa pudica* on the seats of a hackney cab and ordered the cab driver to take a tour of the most interesting parts of the city at a moderate trot and, if possible, without stopping. He did not enjoy

The *Mimosa pudica* is a sensitive plant native to Latin America and the Caribbean that has spread to many countries in the tropics.

the scenery very much, as he was too busy watching for and noting the minutest changes in the behavior of the plants, which had closed their leaves upon the first vibrations of the carriage on the paved streets of Paris.

All things considered, the young student must have thought that it was not a very interesting experiment and that Desfontaines would not be very satisfied. As was to be expected, the plants closed their leaves at the first vibrations of the carriage . . . So? What did his teacher expect from the experiment? Whatever it was, it did not seem to be the right day for a satisfactory result. However, as he continued the tour, something unexpected happened. First one, then two, then five more, and finally all the plants began to open their leaves even though the vibrations of the carriage continued with equal intensity. That was an interesting fact. What was happening? The unknown student had a sudden inspiration and wrote in his notebook: "The plants are getting used to it."

The results of the experiment conducted in the streets of Paris later became an interesting paper for the botanical society and a brief text called *Flore française*, written by Lamarck and Augustin Pyramus de Candolle (1778–1841), but the findings were soon forgotten, as happens to many brilliant insights far more often than we'd like to think. Yet the implications of Desfontaines's test were clear even back then, and they pointed decidedly to an adaptive behavior in plants resulting from the storage of information. How could the *Mimosa pudica* have gotten used to the buffeting of the carriage if it did not have some form of memory? It was certainly a fascinating question, which, however, was long denied a scientific answer.

Then, in May 2013, Monica Gagliano, a researcher from the University of Western Australia in Perth, transferred to my lab in Italy for six months. When she arrived at the LINV (Laboratorio Internazionale di Neurobiologia Vegetale, the international laboratory of plant neurobiology that I run at the University of Florence), Monica was a researcher of marine biology with widely varying interests, ranging from philosophy to the evolution of the species to botany, and the purpose of her visit was to deepen her knowledge of the plant world—or rather, of one particular aspect of the plant world: the behavior of plants. During lengthy discussions about our respective fields of study, we began to plan some experiments that could, on the one hand, justify her period at the LINV in the eyes of her university, and, on the other, provide some answers to at least some of the many questions that emerged during our conversations on the behavior of plants. On the basis of one of those, it seemed essential to me to be able to prove something experimentally that many have believed to be true for some time but without any real scientific basis, which is that plants are equipped with an effective memory. Having agreed on what to research, the most difficult aspect still remained to be resolved: how to demonstrate that plants improve the efficiency of their response thanks to a particular form of memory.

A few months earlier, during a visit to the headquarters of the Japanese LINV in Kitakyūshū, my dear friend and colleague Tomonori Kawano, who runs that section, had shown me, with legitimate pride, some of the many thousands of volumes that the Paris-Sorbonne University had intended to dispose of and that he, thanks to a skillful negotiation, had managed to save from destruction and have shipped to Japan. Amid the many wonders there was an original copy of *Flore française* by Lamarck and Candolle in which they recounted Desfontaines's experiment on the effects of transporting the *Mimosa pudica* plants through the streets of the French capital. So it was that that story of unlikely journeys in carriages came back to my mind, and I spoke to Monica about it. Would it be possible to come up with a remake of this classic experiment, recasting it in such a way as to make it scientifically plausible? After a few days, the new protocol of what we immediately saw fit to call the "Lamarck and Desfontaines experiment" was ready. In 2013 it was unthinkable to reenact the carriage ride with plants, but we wanted to reenact the repetitive stimulation. The purpose of the experiment was twofold: first, to show that the *Mimosa pudica* was able, after a number of repetitions, to identify a stimulus as not dangerous and therefore not to close its leaves; and second, to ensure that the same plant, after a suitable period of preparation, was able to distinguish between two stimuli, one of which was previously known to it, and to respond appropriately. In other words, we were curious to know if the plants would be able to remember a stimulus that they had been subjected to and that was not dangerous, and distinguish it from a new, potentially dangerous stimulus.

In a short time, we prepared a simple but effective experimental apparatus. The new "Lamarck and Desfontaines experiment" required that the plants, placed in jars, be subjected to repeated falls from a height of about 4 inches. The drop, precisely quantifiable, represented the stim-

Properly trained, *Mimosa pudica* seedlings recognize stimuli that are not dangerous (such as a fall of a few inches) and learn not to close their leaves when subjected to them.

ulus. The results were immediately exciting, confirming the correctness of Desfontaines's observations: after a series of repetitions (about seven or eight) the plants began to leave their leaves open, ignoring with sovereign detachment every subsequent fall. Now we had to understand if that were due simply to fatigue or if indeed the plants had understood that there was nothing to fear. The only way to do that was to subject them to a new, different stimulus. So we set up a contraption that would shake the jars horizontally and began to subject the plants to this new stimulus, which was also perfectly quantifiable. The plants reacted by immediately closing their leaves—a fantastic result. Thanks to the "Lamarck and Desfontaines experiment," we managed to prove that plants can learn that a certain stimulus was harmless and distinguish it from other potentially dangerous ones; that they are able to remember a past experience. But how long does this memory last?

In order to answer that question, after leaving about a hundred plants that had been trained to distinguish between the two stimuli undisturbed for a while, we periodically checked to see if they still retained the memory of what they had learned. The result exceeded all expectations: the *Mimosa pudica* remembered for more than forty days—a very long time if compared to the standard memory span of many insects and closer to the standard of several superior animals.

How such a mechanism works in organisms that have no brain, such as plants, is still a bit of a mystery. Extensive research carried out in the field of stress memory seems to demonstrate the fundamental importance of epigenetics in the formation of this type of memory. Epigenetics refers to certain biological mechanisms that activate some of an organism's genes while inhibiting the expression of others. In other words, it is the study of changes that alter the way genes behave but not their underlying sequence.

As has often happened in the history of biology, many advances in understanding have been made possible as a result of plant research, and in recent times this has been especially true of research into the mystery of plant memory. One specific case is how plants remember the exact moment when they must bloom. Their reproductive success and ability to generate progeny are based, above all else, on their ability to bloom at the right time. Many plants bloom for only a set number of days after the end of the winter cold. They are therefore able to remember how much time has passed.

This is clearly an epigenetic memory, but nothing was known about how it worked until recently. In the September 2016 issue of the journal *Cell Reports*, the group coordinated by Karissa Sanbonmatsu of the Los Alamos National Laboratory published the results obtained from studying a particular sequence of RNA that controls the time of flowering of plants in the spring by detecting how long it has been since their exposure to cold. When this piece of RNA, called COOLAIR, is disabled or removed, the plants are unable to bloom. COOLAIR mechanisms could be much more common than we once thought and represent the basis of how plant memory works. Epigenetic modifications seem to

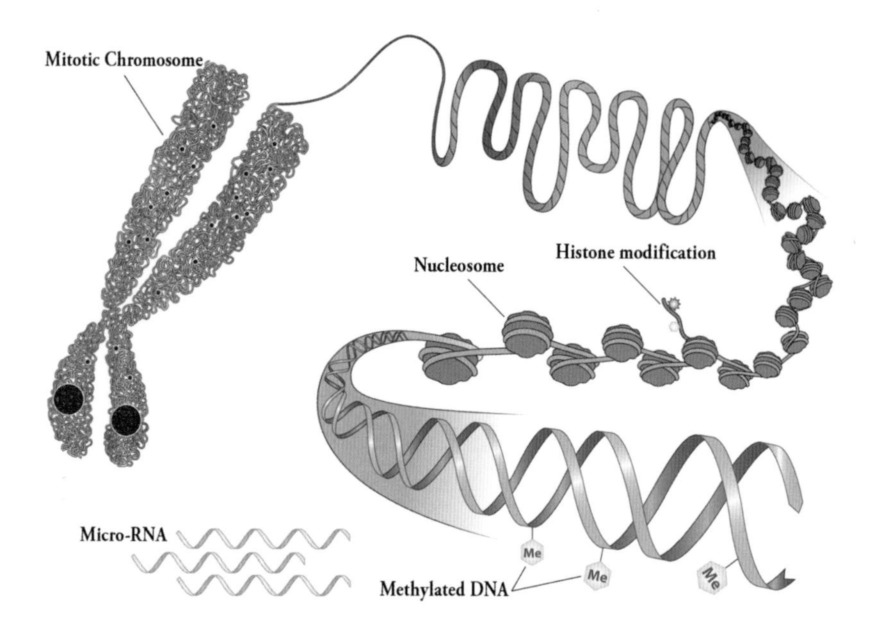

DNA methylation is the most common of epigenetic modifications.

play a greater role in plants than in animals. It is likely that alterations of gene expression following stress may be remembered by cells resulting from epigenetic modifications. Recently, a research group directed by MIT biologist Susan Lindquist advanced a hypothesis: that plants, at least in instances such as the memory of flowering, may use prion proteins, infectious agents that transmit a malformation to all nearby proteins in a kind of domino effect. Prions in animals are not good news; for example, they cause Creutzfeldt-Jakob disease, better known as mad cow disease. In plants, however, they may provide an original form of biochemical memory.

Contrary to what one might think, the importance of these studies goes beyond the purely botanical, even though that is high. In addition to solving the mystery of how plants remember, understanding the workings of memory in organisms without brains will help lead to a better understanding of how human memory works: what mechanisms lead to its alteration or pathologies and how its distinctive forms can be situated outside the nervous system. Any discovery about the biological function of memory also holds great potential for technological applications. Our advances in research on this issue have potential beyond our imagining.

CHAPTER 2

MOVING WITHOUT MUSCLES

Consciousness is only possible through change; change is only possible through movement.

—**ALDOUS HUXLEY, *The Art of Seeing***

I move, therefore I am.

—**HARUKI MURAKAMI, *1Q84***

PREVIOUS PAGE: The dandelion (*Taraxacum officinale*) is a very common species belonging to the Asteraceae family. As can be inferred from its scientific name, it is a very widespread plant whose medicinal properties have been known since antiquity.

AND YET THEY MOVE!

In 1896, just months after the official introduction of cinema by the brothers Auguste and Louis Lumière in Paris, a German botanist named Wilhelm Friedrich Philipp Pfeffer (1845–1920), already in the prime of his scientific career, created the first time-lapse movie. Pfeffer had been working on the technique for many years, ever since he had been privileged to witness the first experimental film in history: the famous stop-motion gallop of a racehorse filmed by the English photographer Eadweard Muybridge in 1878. After that, capturing the movements of plants on film and speeding them up so that everyone could understand their beauty and meaning, but above all as a means to study them, had become a true vocation for Pfeffer.

As a young assistant to the great botanist Julius von Sachs (1832–1897) at the University of Würzburg, he had participated in experiments studying the gravitropism (movements made in response to gravity) of roots. When the experiments proved Sachs's theories to be incorrect and instead corroborated the thinking of Charles Darwin on the issue, Pfeffer's skills as a researcher were called into question by Sachs, and his prospects of continuing his research in Germany were greatly diminished. Even in those days, contradicting a powerful supervising professor did not facilitate a happy university career. Driven by the need to restore his reputation as a researcher, Pfeffer began thinking about how to transform the new techniques of cinema into a tool to study the movement of plants. He realized that in just a few seconds or minutes of film, he could show events that in real time would take hours or days (or even months and years) to happen.

The racehorse Sallie Gardner photographed by Muybridge in 1878. Each image in the sequence corresponds to about one twenty-fifth of a second.

For centuries, biologists and botanists had purposefully avoided conceptualizing this form of plant behavior, trying by every means to safeguard the validity of the time-honored categories of "animals" and "plants" and defining as "anomalies" or "aberrant variations" plants that showed rapid movements. (They had even begun to categorize them as part of the world community of animals or the "zoospore," just to underline their affinity to the animal world.) The surprise and the enjoyment, even today, that anyone experiences when faced for the first time with the quick movements of a plant such as the *Mimosa pudica* are testament to the profound conviction that immobility is the key feature that distinguishes plants from animals.

Pfeffer's attempt to present the motor skills of plants was the first in the history of science to succeed. A few months after the first screening by the Lumière brothers, Pfeffer displayed to an astonished audience of botanists the sensational applications of this new technique. For the first time in history, he showed plants in action, enabling the study of their moves and thus their behaviors. Before the stunned faces of colleagues, the German botanist showed the flowering of a tulip, the diurnal and nyctinastic—sleep—movements of the *Mimosa pudica*, the continuous movement of the Asian shrub *Codariocalyx motorius* (telegraph plant), and finally, the jewel of the collection, the hardest thing to show: the growth and exploratory movements of roots in soil, so similar to those of an ant or earthworm underground.

With the invention of time-lapse photography, Pfeffer gave botanists a tool to render visible what had been invisible. Just as Hans Lippershey's telescope (no, it was not Galileo who invented it) had made possible the study of the infinitely distant universe and Zacharias Janssen's microscope had made the infinitely small observable, so Pfeffer's new cinematographic technique made possible the study of the infinitely slow.

Access to this new dimension of reality was not without its consequences. Plants—in fact, representing almost all of life on earth—which until then had been perceived more as objects than as living beings, began to reveal their mysteries and the disconcerting variety of their movements. It was a real revolution in people's common perceptions. Those who had hitherto looked at a rosebush or lime tree as something aesthetically pleasing but more or less inanimate began to show an interest in and a new respect for plants. It is no coincidence that between the late nineteenth century and the First World War a multitude of studies flourished on tropisms (movements that depend on the direction of a stimulus), on nastic movements (which are independent of external stimuli), on the movements and behaviors of plants generally, and, finally, on plants' cognitive capacities. Those culminated in the opening paper read on September 2, 1908, by Sir Francis Darwin (1848–1925),

The *Codariocalyx motorius* (telegraph plant) is a leguminous plant that is widespread in the tropical areas of Asia. Its special feature is its ability to move its lateral leaves fast enough to be seen with the naked eye. The function of this movement is still unknown.

at the annual conference of the British Association for the Advancement of Science, during which the first professor of plant physiology and son of the great botanist Charles Darwin stated very clearly that plants were organisms endowed with intelligence, reflecting the thoughts of his father, who had written in his 1880 book *The Power of Movement in Plants* that plants are endowed with powers to direct movements that are "like the brain of the lower animals." Pfeffer's ingenuity still allows us to assess and analyze various types of movement in plants, to distinguish between active and passive, and to better understand the mechanisms by which they occur.

CONES AND OAT AWNS

Plants have active movements, which require the consumption of internal energy, and passive ones, which, on the contrary, do not need internal energy and instead use energy present in the environment. For example, many plant organisms exploit the difference in humidity between day and night to perform quite complicated actions. In general, an important feature common to all plant movements is that they are not based on the operation of complicated protein structures such as muscles but are mostly "hydraulic," based essentially on the simple transportation of water, whether in the form of liquid or steam, inbound to and outbound from the tissues.

In the so-called active movements, the creation of movement is the direct consequence of changes in the flow of water through cell membranes. Water, entering the cell, causes an increase in pressure that pushes the membrane against the cell wall and thus induces rigidity in the organ and therefore movement. By actively controlling cellular pressure, plants can generate movements such as the opening of stomata and flowering; the mimosa can close its leaves; and the Venus flytrap can spring its trap.

Passive movements, on the other hand, are due to variations in some of the constituents of the cell wall, which is the skeleton of the plant. There is nothing like this sturdy structure in animal cells. It is the structural component that gives a plant cell its rigidity and the ability to hold its shape, and it is made up of cellulose fibers embedded in a soft matrix of soluble proteins and other substances. It is this soft matrix that, reversibly expanding when it combines with water molecules, is responsible for the opening of cones, the explosive opening of wisteria pods, and the movement along the ground of the *Erodium cicutarium* or wild oat seeds.

The cone—that is, the organ that contains the reproductive structures of coniferous trees—manages, for example, in an enterprise that is anything but simple for dead tissue, to open its woody scales in a dry environment and close them again when the humidity in the air is high. Have you ever seen a pinecone on a rainy day? If so, you will have noticed that it is firmly closed to prevent the seeds escaping, while on a

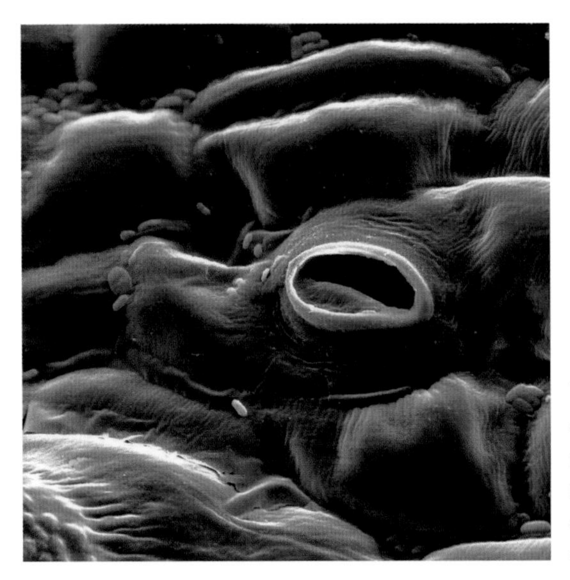

The image of a single tomato stoma under an electron microscope. The CO_2 needed for photosynthesis enters the plant through the stomata.

The *Dionaea muscipula* (referring to Dionaea, the mother of the Greek goddess Aphrodite), also known as the Venus flytrap, is a carnivorous plant native to the swamplands of North and South Carolina (United States).

sunny day the scales open fully to allow their release. Apparently, such a strategy is used because on humid or rainy days, the seeds would fall so close to the parent plant as to prevent their effective dispersal in the environment.

How does this movement, simple in appearance but in fact incredibly complex, actually take place, especially if, as with a cone, it is dead tissue that does not use any internal energy? The trick lies in the nature of the scales. Each of them consists of two different tissues, indistinguishable to the naked eye. Only through careful microscopic observation can their differences be seen. The internal surface of the scale is made of a thick-walled fiber grouped together to form something like microscopic ropes, while the outer surface is composed of variably shaped cells, which are bigger and shorter. The two components have different affinities for water, and therefore it is said that they are hygroscopic in different ways. When water is absorbed or lost from these

The cone (or strobilus) is a structure formed by woody bracts that house the gymnosperm seeds. In the pine tree, the bracts are arranged in spirals according to the Fibonacci series.

fibers, the tissues expand or shrink unevenly, creating a macroscopic closing or opening of the cone.

The phenomenon can be easily replicated in the laboratory (also at home: just dip an open cone into water to see the effects) and has produced a long series of studies, many of them motivated by the desire to create artificial materials that would perform similarly. Imagine how many possible applications you could find for a material capable of moving by making use of only humidity gradients. In 2013 Dr. Mingming Ma and some coworkers at MIT developed a polymer film capable of exchanging water with the environment and expanding and contracting rapidly, thus generating movement: it can build up an extraordinary pressure and lift objects 380 times as heavy as itself. Furthermore, by linking this actuator with an electric element, scientists can produce electrical energy with a peak voltage of approximately 1 volt, enough to charge microelectronic and nanoelectronic devices—all of this using only humidity gradients.

The practical, useful possibilities of all this are many. We, for exam-

The *Avena* genus includes many species native to Europe, Africa, and Asia. Some of them have been cultivated for thousands of years as a source of both human and animal food.

ple, are attempting to exploit a similar system to make the sensors that monitor the electrical activity of trees energy self-sufficient. But it is certainly not the only way forward. Systems of this type (which, I remind you, are almost microscopic in size), distributed in the fibers of our clothes, in upholstery, or in other tissues, would render them energetically autonomous and able to operate any type of sensor or device that is not too energy intensive. We can imagine fabric that, in contact with our bodies, would be able to detect the most important clinical data; or materials capable of measuring environmental parameters, stress levels, or other physical or psychological states. Soon all this will be reality, and a decisive part of the technologies and materials that will enable these advances will have been inspired by how plants work.

A VERY ACTIVE SEED: THE *ERODIUM CICUTARIUM*

Among all passive movements that we can find in the world of plants—and there are some really bizarre ones—none, in my opinion, is more interesting or curious than those carried out by *Erodium cicutarium* seeds, which explode when they come off the parent plant, may be carried away by the fur of some passing animal, fall to the ground, move until they find a crack in the ground, and then, finally, slip into it. Here is a sequence that is truly worthy of note, difficult to carry out even for organs that have internal energy and impossible to imagine of dead tissue.

The *Erodium cicutarium* is a small and pretty plant from the same family as the familiar geranium (Geraniaceae) of our balconies, which grows wild in many parts of the world. Its name is bound to the shape of the fruit, which resembles the beak of a heron (*erodiòs* in ancient Greek), and to the resemblance of its leaves to those of the hemlock. Other genera of the family also have names that recall the beak of waders. The word *geranium* itself derives from the Greek *géranos* (crane), while the *Pelargonium* (the name of a genus of the family) comes from the Greek *pelargòs*, meaning "stork." Coming back, though, to our *Erodium*, it is an annual herbaceous plant, quite widespread, with light purple flowers that have five petals. Its most outstanding feature is definitely its seeds. Each of them is formed by the actual seed, hirsute and sharp like the tip of a spike, and a twisted spiral awn, which is also covered with hairs. Each of these elements has a special function in the surprising series of movements that it can produce.

My interest in the *Erodium* began some time ago, when one of the researchers in my laboratory, Camilla Pandolfi, went to work for a couple of years in a special office of the European Space Agency

The *Erodium cicutarium* (of the Geraniaceae family) is an annual or biennial herbaceous plant native to the Mediterranean basin.

(ESA). This office, with the fascinating and promising name of Ariadna Advanced Concepts Team, was a liaison between ESA and the European academic community interested in advanced research in space technology. When Camilla came to ask me what I thought of her possible transfer to that center, I had no doubt: she had to go. Right away. Two years in a research center with such a promising name would be a fantastic experience. Besides, our laboratory had concentrated for years on the behavior of plants in the absence of gravity and continues to actively cooperate with several space agencies. Camilla would feel at home there.

Once she transferred to ESA's main research and development center in Noordwijk, in the Netherlands, about thirty miles from Amsterdam, Camilla's new assignment turned out to be even more interesting than she had imagined: she was tasked with studying examples of the materials, functions, and strategies of the plant world and then using those studies to provide new perspectives for the advancement of space technology. In a short time she identified a number of topics whose consideration might lead to interesting innovations. Among them, two were also very important for us: the study of tendrils as a model for artificial prehensile organs and research on the *Erodium* seed as a starting point for the construction of probes that can penetrate into extraterrestrial soils and analyze it, using little or no energy.

Pathfinder, *Spirit*, *Opportunity*, and *Curiosity* are the latest in a list of robots sent to explore Mars, along with the newest arrival, *Philae*, which reached the surface of the comet 67P/Churyumov-Gerasimenko on November 12, 2014. For each of them, drilling into the soil and analyzing samples taken at a certain depth were among the main objectives of the mission. The discovery of water even in frozen form, the study of the chemical composition of the soil or even the possible presence of microscopic life, make the drilling of celestial bodies one of the priorities for all space agencies worldwide. A device being sent into space must satisfy thousands of specifications, but above all it has to meet two fundamental requirements: it must weigh as little as possible and consume a minimal amount of energy. Weight and energy are two insurmountable constraints in all kinds of space technology. And that is why, with its light structure and its ability to move across and penetrate the soil without consuming energy, the *Erodium* provided a significant field of research for ESA.

Like all plants, the *Erodium* needs to disperse its seeds on as wide a surface as possible. For instance, a parent plant has no interest in having all its babies around it; on the contrary, it implements every strategy to ensure that they move away from it. There are many good reasons that make this choice evolutionarily relevant, not least to prevent the growth of nearby rivals with which it would have to compete.

Plants have invented hundreds of different solutions to spread their seeds in the environment, ensuring their best chance of survival. In the

In the spring, *Erodium cicutarium* seeds disperse through an explosion thanks to the tensions caused by the fruit changing shape during its maturation.

case of the *Erodium*, everything begins with an explosive movement. The seeds are grouped so as to accumulate mechanical energy as if they were a spring. This energy continues to increase until any disturbance in the balance, such as the slight touch of an insect, the passing of an animal, or even a gust of wind, causes the immediate release and explosion of the seeds. They are literally catapulted as far as several feet away; then, by hooking onto and binding themselves to the fur of animals, they can be taken miles away from the mother plant.

Once on the ground, a new adventure begins. The long bristlelike appendage of the seeds known as the awn (which closely resembles sperm) begins to curl up in reaction to the humidity in the air. The bristles help it to move, and as soon as the seed finds a small slit in the soil, they help move it into exactly the right position with its tip pointing down. At that point, with the spiked apex having been introduced into the slit, the coils caused by the humidity variation between day and night provide the necessary driving force to penetrate the soil. Each winding of the coil making up the awn propels the seed deeper. Furthermore, the shape of the tip ensures that the motion of penetration remains constant, both when the coil winds and when it unwinds. In a matter of days—that is, a few day/night cycles—the seed reaches its final position, many centimeters deep, ready to germinate and develop into a new plant.

Considering the extraordinary potential of *Erodium* seed, it is easy to understand why the research conducted by Camilla and her colleagues at the Advanced Concepts Team kept them busy for almost a year, during which every possible aspect of the forces and strategies carried out by this wonderful plant was investigated in detail. To study the potential in building self-burying probes for use in unmanned planetary exploration missions, they evaluated the capacity of the seeds to dig into various soils with mechanical properties similar to those found on the Moon, Mars, and asteroids.

To examine the numerous movements, it was necessary to use very different video recording techniques. The *Erodium* has both slow movements, which require the time-lapse techniques invented by Pfeffer, and superfast movements, requiring videos that slowed the movements down in order to study them in detail. Analysis of the slower movements, such as digging into the soil, therefore required time-lapse techniques suitable for seeing the twisting cycles resulting from the changes in humidity between day and night. In order to study the explosive ejection of the seeds or their subsequent landing on the ground, high-speed video equipment was employed.

This was by no means a simple task. In our laboratory, we are experts in time-lapse techniques, but we had no idea how to film the fast movements involved in the explosion of the seeds and their flight. Such filming required very different skills and equipment, and it took a while for us to work out how to do it. The problem was that it was not enough to record one or two seed explosions; we had to be able to film literally thousands, under different conditions of humidity and temperature and on different soils that mimicked the possible extraterrestrial soils. We also needed a system to trigger the seeds on command when all the conditions were ready for the experiment.

We spent over a month without finding any solution more effective than starting the recording and waiting until the plant decided to "explode." By recording a thousand frames per second in HD resolution, the amount of data that we gathered for even a few minutes of footage was hyperbolical (many gigabytes per second). We did not have any equipment capable of holding hours of recording time. In short, we were in trouble. We had captured only a handful of explosions in one month, and we were lacking ideas about how to proceed until, one auspicious day, the solution appeared in the guise of a middle school boy visiting our lab. Before entering the LINV, all visitors—no matter their age—are by law given a brief lesson in conduct, which includes being explicitly asked not to touch anything. The prohibition serves both to prevent damaging delicate instruments or experiments in progress and to prevent visitors from being harmed. Fortunately, that day, one of the students saw fit to disobey.

Upon reaching the equipment that we used for the experiments with the *Erodium*, and while one of my staff was explaining the peculiarity of that plant, a young boy exclaimed, "The *Erodium* is so cool!" Then he pulled a thin wooden stick from his pocket and touched the seeds that were still on the plant right at their intersection, causing their immediate expulsion. While the professor who was accompanying them apologized for the unruly schoolboy, promising that he would be disciplined, I was struck by the result of that simple gesture. The young student, who came from an area not far from Florence that was rich in spontaneous *Erodiums*, had learned, while playing in the meadows, how to set off the explosion. All that was needed was a light touch where the seeds are in contact with one another for the elastic force that binds

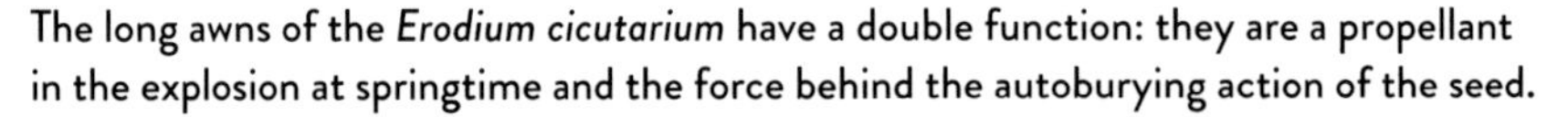

The long awns of the *Erodium cicutarium* have a double function: they are a propellant in the explosion at springtime and the force behind the autoburying action of the seed.

them to be released. Finally we had a practical system to induce the expulsion and could therefore proceed with our studies. In the following months, we carried out thousands of "controlled explosions." Thank God for rebellious children.

The results obtained at the end of the study prove that every single aspect of the *Erodium* seed has its precise function. The seed's ability to perforate the ground and bury itself is in fact tied to:

1. The structure of the seed
2. The structure of the awn and its movement in relation to humidity
3. An inactive area of the awn
4. The barbs on the carpel and awn

The data collected were used to construct a model of the movement of the *Erodium*, which we delivered to ESA together with a voluminous document that described in great detail the features of this fascinating plant. A future probe for space exploration inspired by the *Erodium* is a real possibility.

CHAPTER 3

THE SUBLIME ART OF MIMESIS

The more I study nature, the more I become impressed with ever-increasing force, that the contrivances and beautiful adaptations slowly acquired through each part occasionally varying in a slight degree but in many ways, with the preservation of those variations which were beneficial to the organism under complex and ever-varying conditions of life, transcend in an incomparable manner the contrivances and adaptations which the most fertile imagination of man could invent.

—**CHARLES DARWIN,** ***On the Various Contrivances by Which British and Foreign Orchids Are Fertilised by Insects***

PREVIOUS PAGE: The plants belonging to the *Lithops* genus are commonly known as living stones. They grow in the driest areas of South Africa and Namibia.

A MODEL, A MIMIC, AN OPERATOR

When we think of mimicry in nature, the examples generally belong only to the animal world: the chameleon, the stick insect, the praying mantis, a number of butterflies, caterpillars, fish, snakes, and many other species. Yet some plants can compete on an equal footing with the most daring animal mimics and indeed in many cases reach levels of mimicry that surpass those of animals.

There are several forms of mimesis in nature that fall into two categories: mimicry (from the Greek *mimetikos*, meaning "imitative"), when one organism imitates another in behavior, shape, or color; and camouflage (or *crypsis*, from the Greek *kryptòs*, meaning "hidden"), which is when an organism becomes invisible by imitating the surrounding environment.

Any organism, regardless of its level of complexity, must have the ability to maintain its internal organization in the face of the natural drive toward degradation and disorder. This ability manifests itself in the skill of being selective and knowing how to make the right choices, for example, distinguishing friends from enemies or expanding or contracting depending on the availability of resources. Every living organism exchanges data with its surrounding environment that allow it to survive. Communication is an essential feature of life; without it, even the simplest organisms cannot maintain the very delicate balance that is life itself.

An organism must respond at all times to the need for recognizing objects, other members of its own species, and all manner of dangers. Interaction with other organisms in certain phases of the life cycle is an unavoidable necessity.

When one living being emits a signal of any kind (visual, olfactory, auditory) to another, in order to influence its behavior in its own favor, it is a mimetic phenomenon. Because there is mimicry, there must therefore be a model (that is, the issuing organism that produces the authentic message), a mimic (which benefits from reproducing the model's signal), and, finally, an operator (which must respond to the message in a way that is useful to the mimic).

THE *BOQUILA TRIFOLIOLATA*, KING OF MIMESIS

For me it is in the plant world where the sublime art of mimicry reaches levels of bravura that have no equal among animals. The study of the refinement of the mimetic abilities plants can achieve can help us understand the unexpected sensation capacity of plants. This is the case with the *Boquila trifoliolata* and its extraordinary mimetic ability.

The *Boquila* is a veritable Zelig of the plant world; certainly the most extraordinary example of mimesis that you can find in nature. It is a liana, a woody, ground-rooted vine that grows in the temperate rain forests of Chile and Argentina and has the distinction of being the only species of its genus. In reality, it is a rather common plant, so much so that it is known by several names in Chile (*pilpil, voqui, voquicillo, voquillo, voqui blanco*), and it produces edible berries. The species has been known for a very long time and has been studied by hundreds of botanists, experts, and admirers who have seen it grow and prosper in its original habitat. Yet, until a few years ago, no one had noticed its incredible mimetic capability. In 2013, during a quiet walk in a forest in southern Chile, the botanist Ernesto Gianoli came across the *Boquila trifoliolata* for the umpteenth time in his career. Nothing strange there, the plant was already well known and described, but this time something caught his attention. A botanist in the forest is like a collector at a flea market: senses on the alert, looking for something that everyone else has missed. When botanists wander in the forest to find new species, their eyes are trained to see the most hidden details, every little difference in shape or color, something new that differentiates the plant that they are studying, including details of secondary importance. So in observing a very common shrub more carefully in that region of Chile, Ernesto noticed some variations in the shrub's leaves. Looking closely, he noticed that they did not belong to the shrub in question, but to a climber that was growing around it. The climber was a *Boquila trifoliolata*, but its leaves were strikingly similar to those of the shrub it was climbing.

Ernesto looked around curiously to see if any other *Boquila* plant nearby had the same characteristics. What he found left him speechless: on every shrub or tree around which it grew, the *Boquila trifoliolata* mimicked the leaves of the "host" species every time, and with great skill. Not only that, but it seemed capable of replicating the most diverse leaves. As far as he knew, no other plant was capable of doing anything similar; even orchids, which are considered to be the champions of plant mimesis, are able to imitate only one species or at the most produce flowers that resemble those of many different species. The capacity to imitate different models had until then belonged exclusively to the animal world. Galvanized by the discovery but also a bit incredulous of what he had seen, he and his student Fernando Carrasco-Urra began the long series of tests and checks required to render his discovery unassailable. It would not be easy, in fact, to convince the scientific community that a plant is able to mimic the shape, size, and color of completely different species. In the end, the result was even more amazing than he could have imagined.

Not only is the *Boquila* able to imitate the many species it climbs, it actually does much more. Growing in the vicinity of two or even three different plant species, a single plant can change its leaves so that they are confused with those nearest to it every time. In other words,

The *Boquila trifoliolata*, a liana plant that is very common in the temperate forests of Chile, has incredible mimetic capabilities. In the image we see the leaves in their normal condition.

the *Boquila* can change the shape, size, and color of its leaves several times, depending on which species grows closest to it. Gianoli and Carrasco-Urra's discovery has huge consequences. To be able to regulate the features of its leaves with such flexibility means that the plant is modulating the expression of its genes in a way never seen before.

The *Boquila* is a unique case of mimicry, and I can say with some confidence that there is no other example of mimicry that involves near-simultaneous changing of the shape, size, and color of the organism's body. Only one of these transformations—color—is frequently seen, and although two are sometimes exhibited, all three together is something new and a phenomenon not found even in the animal world.

The mimetic capacity must somehow be of advantage to the mimic, in this case to our *Boquila*. But what is the benefit for a plant that modifies its leaves to imitate those of its host? One possibility is protection from harmful insects. If, for example, the leaves the *Boquila* mimics belong to a plant that is poisonous for herbivorous insects and thus the latter have learned to avoid them, the mimic, in being mistaken for them, gains an advantage. This form of mimicry is called *Batesian*, after the English naturalist Henry Walter Bates (1825–1892), and refers to cases when a sheep disguises itself as a wolf, so to speak. Examples of Batesian mimicry in the plant world are quite common, and some of the better-known ones include protection from herbivorous animals that certain species of the labiate (mint) family have achieved thanks to their ability to imitate stinging nettle leaves almost to perfection. A second hypothesis, which is simpler and therefore preferable, is that by mixing its leaves with those of another plant, the chances of being attacked by herbivorous insects statistically diminish for our *Boquila trifoliolata*. In fact, in the case of an assault, the host, with its more numerous leaves, will be more damaged than the *Boquila*. We still do not know which is the correct theory. It is likely, however, and as is often the case, that the explanation will involve a combination of factors.

Ernesto Gianoli tells us that the specific features of some leaves, such as serration (sawtooth edges), are difficult for the *Boquila* to reproduce, yet it was evident that the plant "did its best to emulate them," generating leaves with "draft" serration along the edges. Observing the phenomenon, however, is only the beginning of the story. The most important question raised by the *Boquila*'s behavior is not so much how

The *Stachys sylvatica*, more commonly known as hedge woundwort, perfectly imitates nettle leaves.

The common nettle (*Urtica dioica*) has leaves and stems covered with a severely stinging substance that it uses for defensive purposes.

it can modify its body with such speed but rather how it knows what to imitate. In the study that describes for the first time the extraordinary mimetic characteristics of this species, Ernesto Gianoli and Fernando Carrasco-Urra offer two hypotheses. The first is that because of its perception of emissions of volatile substances, the *Boquila* plant is able to identify the model to imitate. But this is a highly unlikely supposition since the *Boquila* imitates the leaves that are closest to it even when surrounded by a blend of volatile compounds produced by dozens of different species. The second hypothesis, which assumes a possible horizontal transfer of genes from the host plant to the *Boquila*, carried by some microorganism, seems even more improbable.

In September 2016, I—together with my dear friend and coworker

Professor František Baluška from the University of Bonn (we have written about fifty scientific papers together)—offered a new solution to the puzzle: that the plant has some sort of visual capacity. It may sound like an incredible hypothesis or even science fiction, but it seems to me the one that is most likely to be true.

As early as 1905, the famous Austrian botanist Gottlieb Haberlandt (1854–1945), in one of his writings that at the time caused a sensation in the scientific community and beyond, proposed that plants were able to perceive images—and thus had a kind of visual capacity—due to the cells of their epidermis. Very often, in fact, the epidermis of a plant is convex like a lens and could conceivably convey the images of surrounding plants to the underlying cellular layer. According to Haberlandt, the epidermic cells of plants work like ocelli (a type of minute, simple, primitive eye) found in many invertebrates. Francis Darwin liked Haberlandt's theory and talked extensively about it in his writings on the perceptive capacities of plants, adding to its scientific legitimacy.

At the congress in Dublin where Francis Darwin argued that plants were able to remember and act according to those memories, the British botanist Harold Wager (1862–1929), a fellow of the Royal Society, showed an astonished audience numerous photographs produced using the epidermic cells of leaves of different species as lenses: detailed enough images of people and views of the English countryside that demonstrated, at least from the point of view of simple optics, how the phenomenon of vision in plants was perfectly plausible. Then silence. As happens to numerous theories in biology, especially those concerning plants, Haberlandt's was forgotten. Nobody went to the trouble of seeking further evidence to confirm it or to deny it altogether. Visual capacity in plants was apparently too eccentric an idea to be taken seriously.

Haberlandt's theory fell into oblivion, and no scientific article has mentioned it in the past century. However, a number of surprising discoveries in the last five years have demonstrated, with strong evidence,

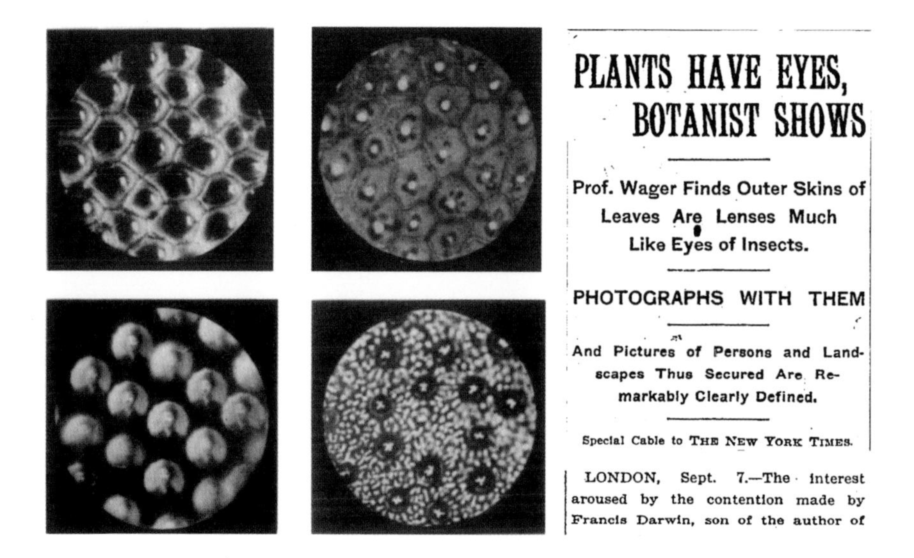
PLANTS HAVE EYES, BOTANIST SHOWS

Prof. Wager Finds Outer Skins of Leaves Are Lenses Much Like Eyes of Insects.

PHOTOGRAPHS WITH THEM

And Pictures of Persons and Landscapes Thus Secured Are Remarkably Clearly Defined.

Special Cable to THE NEW YORK TIMES.

LONDON, Sept. 7.—The interest aroused by the contention made by Francis Darwin, son of the author of

Photographs by Harold Wager using the epidermis of the leaves of plants as lenses. Next to it, the article that appeared in the *New York Times*.

the visual capacity of even unicellular organisms, bringing the theory back into favor and pushing us to a new and deeper reflection on the possibility that vision in plants and Haberlandt's ocelli are something more than simply a charming old theory.

That plants have a rudimentary form of vision is in fact the most plausible explanation for the changing mimetic behavior of the *Boquila*. What is required of the plant to achieve mimesis is something that many single-celled organisms have. A recent study of a prokaryote shows that it is able to measure the intensity and color of light through different photoreceptors and to operate the single cell of which it is composed as a microlens measuring its position relative to a light source. The image of the light source enters through the convex membrane of the cell and is projected onto its opposite face.

Other single-cell organisms—but this time eukaryotes and therefore at a higher level of cellular complexity—have extraordinarily ingenious ocelloids, which work thanks to structures similar to lenses and the retina. Many invertebrates have ocelloids, ocelli, and a whole arsenal of more or less complex organs, though always very different from our eyes; in fact, although we are immediately inclined to think of eyes when it comes to vision, visual systems in nature are many and various. It is quite clear that all the prerequisites exist for some plants to be able to exercise a primitive form of vision.

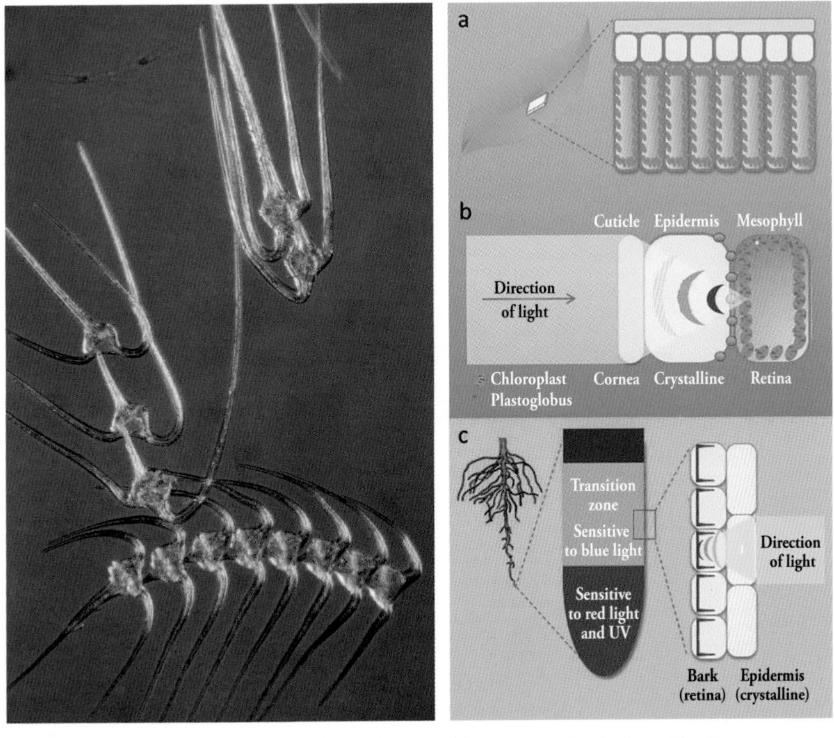

Dinoflagellates are microscopic algae that represent one of the most important groups of phytoplankton. Some species have very complex ocelli.

Features typical of ocelli, with structures similar to those of the cornea and retina, can also be found in the epidermis of leaves and roots.

PLANTS, STONES, AND COLORFUL SIGNALS

Mimics are found everywhere in the plant world. Not all of them are capable of hyperbolic performances like the *Boquila*'s, but they are always fascinating. The cryptic capablities of the *Lithops*, or living stone, is another example. *Lithops* (from the Greek *lithos*, stone, and *opsis*, appearance) is a genus of the Aizoaceae family, which includes species of plants native primarily to the desert areas of Namibia and South Africa. These are plants that, as their name implies, resemble stones. In addition to their mimetic capacity and dexterity, they possess another series of extraordinary adaptations that enable them to survive in their native deserts.

The *Blossfeldia liliputana* is a cactus that, like the Lithops, mimics stones to escape predation.

Lithops aucampiae specimens. The two halves of each oval are the leaves of the plant, which have adapted to conditions of extreme aridity by becoming thick and juicy.

Lithops are very small and have only two leaves, divided by a fissure from which sprout flowers. Succulent and variable in color (from green to rusty red, from cream to grey and purple, through the presence of streaks and stains), the leaves perfectly imitate the shapes and shades of small stones. You have probably seen them in some market, labeled as "living stones" or something similar. To survive the high temperatures and lack of water, these plants have developed a stem that is reduced to the minimum, often underground, so that only the leaves, and often only their flat top, are exposed, thus imitating pebbles in every respect. The *Lithops*' leaves are often windowed, that is, due to the lack of chlorophyll, they have transparent areas that allow light to reach deep into the innermost parts of the plant that are not directly illuminated.

The advantage of being able to blend into the stony desert background is evident for plants that, having no defensive thorns or other armor, would have no chance otherwise of surviving animal predation. In addition, the chromatic combinations and changes in *Lithops* plants are effective communication tools. Although the function of color in evolution is an important field of study in animal biology, in the plant world it has been shamefully ignored, with the exception of the importance of flowers in plant-pollinator relationships. In fact, although plants such as *Lithops* use their shapes and colors to hide from predators, there are other much more common plants that use their shapes and colors to send messages of the opposite kind: to announce their strength or their dangerousness.

One of the most interesting examples of this type of mimicry would seem to be that produced by many arboreal species with their spectacular coloring during the autumn (I use the hypothetical form because it is not yet clear how well founded this theory is). Until a few years ago it was believed that the autumnal explosion of reds, oranges, and yellows coloring our forests was a trivial side effect of chlorophyll degradation, the process of which reveals other colors that until then were masked by the green. The suspicion that the phenomenon instead represents something more complicated came with the discovery that some species invest significant energy in the production of the molecules necessary to color the leaves; and this just a few days, or weeks, before losing them in the autumn. Why invest in something that is so patently useless and that would in any case be lost very soon? The theory formulated by William Hamilton (1936–2000) of Oxford University in 2000 could explain the mystery. According to Hamilton, deciduous trees make an effort to produce brilliant autumn coloring in order to emit a so-called honest signal, that is, a message of strength directed at aphids. And the more obvious the effort to emit this message, the more reliable it is.

Perhaps you're familiar with certain kinds of gazelles that, when they see a lion, start jumping on the spot like a spring instead of trying to escape. At first glance their behavior, too, would seem pointless, a waste of energy. But in fact what they are doing is sending a message to the lion: "See how sturdy and strong I am? It would be a waste of your energy and time to try to catch me." With their intense coloring, trees send a signal of strength and vigor to aphids, which have a migratory peak in the autumn, urging them to seek a more pleasant host elsewhere. It is no coincidence, then, that maple trees—notoriously susceptible to aphid attacks—show some of the most extraordinary autumn colors. Other examples of this type of signal are the peacock's tail and humans' show of various status symbols, both of which are inexplicable unless seen as messages of power. The American scholar of physiology and evolutionary biology Jared Diamond has even argued that some of our excessively risky behaviors, such as bungee jumping, can be classified as signals of this nature.

HUMAN RESOURCES; OR RATHER, HUMANS AS A RESOURCE FOR PLANTS

Agriculture began between 12,000 and 15,000 years ago in the region of the world formerly known as the Fertile Crescent (encompassing present-day Iraq, Syria, Lebanon, Cyprus, Jordan, Israel, Palestine, Egypt, the southeastern fringe of Turkey, and the western fringes of Iran), a region now ravaged by near-constant wars. With agriculture came civilization. When humans abandoned hunter-gatherer activities and became sedentary, cultivating the land, a history of coevolution with plants also began. Some plants became inseparable companions, providing the food needed in exchange for protection and care; but, above all, plants gained a superefficient carrier capable of spreading them across the planet. This was a very good deal for both parties that has worked and continues to work thousands of years after it began. The deal is so beneficial that today three plant species—wheat, corn, and rice—alone provide about 60 percent of the calories consumed by humans, and in return they have colonized huge surfaces on every continent, outclassing any plant competitor in terms of its spread across the earth. The relationship between us and these plants is so close that it can be regarded as a genuine symbiosis. Sixty-nine percent of the carbon of which the average US citizen is composed comes from only one species: corn. But although having a de facto monopoly on human food provision may be a good arrangement for these three plants, tying our own survival to just three or four species is not an advantage for humanity; it is too small a selection of "suppliers" to provide almost our entire caloric consumption.

Moreover, the selection is constantly decreasing. In the past, humanity drew on a much greater number of species. In the eighteenth century, despite there being far fewer edible plants available in Europe than today (all the exotic and colonial plants were not yet available), the number of plant species commonly eaten was three times what it is today. In more ancient times, before the invention of agriculture, humans consumed literally hundreds of different plant species. Over the past ten thousand years, and with a dramatic acceleration in the last century, we have tied our lives more and more to a reduced number of plant species. The lower the number of species we depend on, the higher the risk that something could go wrong. A disease that attacks corn or rice, for example, would be catastrophic for mankind. It has already happened—remember Ireland's Great Famine.

Such a profitable bargain for plants (a huge demand and a consequently growing supply) must necessarily invoke imitators, fraudsters, or simply other players intent on getting their share of the profit. In fact, that is precisely what has happened, and through tricks and deception, many species have attempted to pose as cultivated plants in order to obtain the same benefits. The trick used by many of them is a mimetic phenomenon: transforming their distinguishing features in order to deceive humans.

Since the beginning of agriculture, humanity has chosen plants that, for one reason or another, it considered the best: size of fruit or seed, shape, color, disease resistance, height . . . But each time people decided what features to select for cultivation, the so-called infesting plants learned to respond to those preferences. One of the best-known cases in which a species succeeded in blending with another, so as to receive the benefits of cultivation, is that of the *Vicia sativa* (common vetch) with the *Lens culinaris* (lentil). The latter is one of the oldest species cultivated by man. Its consumption was documented fifteen thousand years ago, and since then it has been one of the most common crops in the Mediterranean area, as seen in the story of Esau in Genesis 25:29–34, when Esau traded his birthright to his younger brother Jacob for a plate of lentils.

The *Vicia sativa* has the same soil and climate needs as the lentil. That made it inevitable that fields of lentils would also contain vetch

The lentil (*Lens culinaris*) was one of the first species domesticated by humans. Archaeological sites attest to its consumption between 13,000 and 10,000 BC.

The common vetch (*Vicia sativa*) is a widespread forage plant that produces seeds similar to the seeds of lentils.

plants and would always do so. But that was not a problem: its round seeds, very different from those of the *Lens culinaris*, could be easily removed. There was no possibility of error. The vetch plant, obviously, did not want to be discarded in this way. So through generation after generation of cultivation, vetch seeds began to change, bringing them closer and closer to the lentil until their seeds become so similar in shape, size, and color as not to be easily distinguished. The deal was done: since it resembled the lentil, humans selected it along with the latter and in that way carried it through to each subsequent cultivation—a very profitable ploy for the vetch, which thus could align itself with the lentil and benefit from all the advantages of cultivation.

This particular type of mimicry, exclusive to the plant world, is called *Vavilovian mimicry*, after the great Russian agronomist and

There are hundreds of varieties of lentils; their seeds have colors ranging from green to brown to yellow to orange.

geneticist Nikolai Ivanovich Vavilov (1887–1943), who first studied it, noting the potential consequences that could ensue. The author of a pioneering paper on the origin and geography of cultivated plants ("Genetics and Agronomy," 1912), Vavilov not only discovered the centers of origin of the cultivated species but also advocated the need to safely store genetically pure seeds. After the first seed bank, still operational in Saint Petersburg, was established, Vavilov's idea had its most important realization in the Global Seed Vault, a facility on the far northwest Norwegian Svalbard archipelago whose aim is to provide a security and preservation net against accidental loss of the traditional genetic heritage of the most important species such as rice, corn, wheat, potatoes, apples, cassava, taro, coconut, and so on, thereby ensuring genetic diversity. The validity and importance of Vavilov's foresight on the need to conserve the seeds of plant species, preserving at least a small part from destruction, have been definitively proven in recent months, when Syria applied to the Global Seed Vault for seeds required to restart farming practices in regions ravaged by war. Above all, however, Vavilov was the first relentless believer that cultivated plants could be improved through genetics, even up to the point of making them capable of flourishing in the most extreme climatic conditions, such as those of certain regions of Russia.

This giant of agricultural science and genetics, sentenced to starve to death in prison by Josef Stalin (ironically as a scapegoat for the Soviet famine of World War II), is now completely forgotten. But even more inconceivable is the fact that his nemesis, the insignificant and awful would-be scientist Trofim Denisovich Lysenko (1898–1976), a supporter of the absurd idea that genetics had no scientific basis and was just a "bourgeois" theory, is today much better known than Vavilov, even among experts.

Vavilov knew a lot, an awful lot, about cultivated plants and related activities. He was the first to observe that man's selection of specific features could induce the mimetic phenomena in other plants, with

The Global Seed Vault can be found on the Norwegian island of Spitsbergen. Its task is to provide a safety and preservation net against accidental loss of the traditional genetic heritage of cultivated plants.

unforeseeable consequences that are not always as negative as one might think. On the contrary, many plants cultivated today exist as a result of this mimetic ability.

The widely grown grain rye (*Secale cereale*), a species cultivated for at least three thousand years, was originally a weed that grew around wheat and barley, with which it shared some key seed features. To understand how a weed such as rye could become a cultivated plant, we must put ourselves into the shoes of the first farmers. Imagine our ancestors, who were slowly abandoning a life based on hunting and gathering and searching for plants to domesticate. Which would they have found desirable? What characteristics were they looking for? They would certainly have opted for species with very large seeds, even better if they were to be found in large numbers inside something that could easily be collected, such as ears of wheat. And certainly they would not have liked plants that dispersed their seeds spontaneously: too laborious to pick them up off the ground. The transformation of humans from hunters to farmers was long and difficult, full of errors and reappraisals. We are sure that some plants such as wheat or barley, with their large seeds and the ears holding them, were perfect for the needs of the first farmers and among the first to be chosen for domestication. Along with those glorious cereals, however, farmers selected their most formidable weeds without noticing they were there.

That is where the history of rye begins, in the unenviable role of weed. The ancestors of the rye that we know today were a classic example of Vavilovian mimicry. Since they were very similar to wheat and barley, to eliminate them, the ancient populations of the Fertile Crescent would have had to carefully search their seeds for intruders. That would not have been an easy task, and as a result, rye became one of the main weeds. When wheat and barley cultivation was expanded to regions further north, east, and west of the original area, rye went along for the ride (man is a superefficient carrier, do not forget), also expanding its own distribution area. Having arrived in regions with colder winters or poorer soils, rye proved its robustness by producing more and better crops than the wheat and barley it had attached itself to, and in a short time it superseded them. Rye had become a domesticated plant.

If the story of the *Secale cereale* is one of Vavilovian mimicry with a happy ending, many others, on the contrary, are not. I am referring to the resistance to herbicides that many species develop in response to the ever higher amounts used in agriculture. In recent decades, the use of herbicides has grown exponentially. And together with an increase that could be termed "physiological," the use of some herbicides, such as glyphosate, has increased pathologically, in part due to introducing the cultivation of plants that have been genetically modified to resist them.

Rye (*Secale cereale*) is a cereal typical of temperate areas. Its cultivation spread about three thousand years ago, migrating from Turkey along with grain, which it mimicked (Vavilovian mimicry).

Amaranth (*Amaranthus palmeri*), a pseudocereal, is a plant native to Central America. Its seeds are edible and similar to those of cereals.

These are specimens on which, for example, glyphosate has no effect and hence prompt farmers to an indiscriminate use of the herbicide. After all, if the main crop is protected and does not suffer any damage, what prevents farmers from using increasing amounts of herbicide until they eliminate all of the annoying "weeds"? The data on the use of this weed killer are disturbing proof of this: in 1974, in the United States alone, 400 tons of glyphosate were used in agriculture; in 2014 that figure had risen to 125 thousand tons. In forty years, therefore, the use of this herbicide increased by more than three hundred times.

This enormous chemical pressure on weeds has fostered the evolution of resistance even in species normally associated with the main crop, that is, mimetic weeds. In the United States today there are populations of *Amaranthus palmeri*—which is an edible cereal but disliked by farmers because it is a weed—that are utterly resistant to glyphosate. The spread of these grains, which thrive in fields of corn or soy, has become a serious problem, and it is being tackled with ever higher doses of glyphosate and with a mix of other herbicides.

Thus pesticide-resistant weeds are increasing everywhere. But I'm not losing too much sleep over this. I have always loved weeds and been fascinated by their ability to survive where they are not wanted and by their intelligence and adaptability. The expansion should not be countered by administering more herbicides, thus destroying all hope of saving our agricultural ecosystems, but with other more environmentally friendly techniques. The damage caused to the environment while we try to stop the proliferation of weeds is far greater than any they could ever cause our crops—always assuming that the crops continue to exist.

CHAPTER 4

GREEN DEMOCRACIES

The class system and authority, being clear violations of the Laws of Nature, are to be abolished. The pyramid of God, King, upper classes and plebs will all be made equal.

—**CARLO PISACANE,** ***Revolution: An Alternative Answer to the Italian Question***

PREVIOUS PAGE: From the nervations on the leaves to the structure of the root system, every component of a plant is in the form of a net.

SOME PRELIMINARY CONSIDERATIONS OF THE PLANT ORGANISM

A plant is not an animal. Even if this statement may appear to be the quintessence of banality, I have found that it is always useful to be reminded of it. In fact, the only idea we have of complex and intelligent life corresponds to that of the animal. And because we do not find traditional animal traits in plants, we instinctively categorize them as passive beings (as "vegetable matter"; and indeed the term *vegetable* is used informally for a person who is so severely impaired mentally or physically as to be largely incapable of conscious responses or activity), denying them any capacity typical of animals, from movement to cognition. That is why, when looking at any plant, we must always remember that we are observing something built on a completely different model from that of an animal—a template so different that, by comparison, all the alien life forms in sci-fi movies are but lighthearted fantasies dreamt up by children.

Plants have nothing in common with us; they are different organisms, a life-form whose last common ancestor with animals dates back to six hundred million years ago, a time when, emerging from the water, life conquered land; when plants and animals went their separate ways, taking different paths. While animals developed the ability to move about on land, plants adapted to the new environment by remaining rooted to the ground and using the inexhaustible light produced by the sun as an energy source. Judging from their success, never has there been a happier choice: today there is no environment on our planet that is not colonized by plants, and their share of the total number of living beings is prodigious. There are different estimates—quite variable, as it is not easy to judge the weight of a life—on the amount of plant biomass on Earth, but no estimate is less than 80 percent. That is, at least 80 percent of the weight of all that lives on Earth consists of plants, a statistic that is a measure, unique and unquestionable, of their extraordinary capacity for success.

The fact that plants from the beginning remained anchored to the ground influenced their every subsequent transformation. They evolved using solutions that were so different from those of animals as to be almost incomprehensible to us. The end result is that plants do not have a face, limbs, or, in general, any recognizable structure that is similar to those of animals, and this makes them virtually invisible to us. We consider them a mere part of the landscape; we see what we understand, and we understand only what is similar to us. That is where the otherness of plants comes from.

What are the characteristics of plants that make them so different and incomprehensible? The first, huge difference is that, unlike animals, they do not have single or double organs that are responsible for the main functions of the organism. For plants, rooted to the ground, surviving the attacks of predators is a big problem: because they cannot escape as an animal would, the only way to survive is to resist predation, to not succumb to it. This is easy to say but very difficult to do. To accomplish this miracle, it is necessary to be built differently from an animal. Plants must have no obvious weaknesses, or at least far fewer than animals do. Organs are points of weakness. If a plant had a brain, two lungs, a liver, two kidneys, and so on, it would be destined to succumb to predators—even tiny ones, such as bugs—because an attack on any one of its vital organs would impair the plant's function. That is why plants do not possess the same organs as animals—not because, as you might think, they are unable to perform the same functions. If plants had eyes, ears, a brain, and lungs, we would not question whether they could see, hear, evaluate, or breathe. Since they do not possess such organs, an effort of imagination is required to understand their sophisticated capabilities.

As we have seen, plants distribute over their entire body the functions that animals concentrate in specific organs. Decentralization is the key. We have discovered that plants breathe with their whole body, see with their whole body, feel with their whole body, and evaluate with their whole body. Spreading each function over the entire organism as much as possible is the only way to survive predation, and plants can do it so well that they can even withstand removal of much of their body without losing functionality. Just as a plant does not have organs on which it depends, it does not have a brain that acts as a central control.

A classic example of plants' staying power is their ability to survive fire. In fact, even against fire, the ultimate destroyer, they have come up with brilliant survival strategies. There are plants that tolerate flames; some are resistant, others have even linked their life and reproduction cycles to recurrent brushwood fires. In all these cases, the ability to survive the destructive power of fire has something of the miraculous in it.

The dwarf palms (*Chamaerops humilis*) found where I spend the summer holidays in western Sicily provide a good example. For as long as I have been going there, extensive fires have frequently ravaged the beautiful hills overlooking the sea, hills covered in flourishing populations of dwarf palms that grow spontaneously, the only palm that is European in origin. Such destruction takes place on average every two years, with surprising regularity (it seems that arsonists observe a strict schedule). Despite this periodic disaster, to which I simply cannot become accustomed, the palms are still there after the fire has been put out. Some are burnt, others reduced to charcoal, others even incinerated. But in a few days, with a humility you would expect from a plant with such a name, they begin to produce new shoots. Heartwarming shiny green shoots that seem even more emerald green against that expanse of black ash appear here and there, popping up from plants that

Dwarf palms (*Chamaerops*, from the Greek *chamái*, on the ground, and *rhōps*, bush) are a widespread species in the Mediterranean maquis.

you could never have imagined were still alive. It is an obvious demonstration of resistance to adversity, a result of the different structure of plants, a structure that is unmatched in the animal world and made possible precisely because of the absence of a command center in favor of a distribution of its functions.

THERE ARE THOSE THAT SOLVE PROBLEMS AND THOSE THAT AVOID THEM

As we've seen, many of the survival solutions developed by plants are the exact opposite of those developed by the animal world. The most decisive divergence between animals and plants is that between concentration and diffusion.

Certainly, the system of centralization typical of the animal structure guarantees a faster decision-making process. However, although responding promptly may in many cases be an advantage for an animal (though, it must be noted, not for all: well-thought-out responses always require time), speed is an incidental factor in a plant's life. What is really important for plants is not so much responding quickly but responding well, so as to solve the problem. At first it might seem rash or even unreasonable to argue that plants find better solutions than animals. Yet if you study the question carefully, you will find that animals respond to the most diverse stresses using the same solution every time, a kind of knee-jerk reaction to all emergencies. This reaction has a name: movement. It is a powerful response, like a wonderful card that trumps everything. Whatever the problem, animals resolve it by moving. If there is no nourishment, they go to where it can be found. If the weather gets too hot, too cold, too wet, or too dry, they migrate to where conditions are better suited. If competitors increase in number or become more aggressive, they move to new territories. If there is no partner with which to reproduce, they move in search of one. The list of possibilities is long, even a thousand different emergencies long, to which there is always only one solution: escape. But escape is not a solution; at most it is a way of sidestepping a problem. Animals, therefore, do not solve problems, they simply avoid them more efficiently.

Since movement is a crucial resource for animals, evolution has worked tirelessly, for hundreds of millions of years, to refine this capacity so that it works in the best possible way, quickly and smoothly. With this in mind, a hierarchical organization of the body with a central command responsible for every decision is the best form any animal could hope for.

For plants, on the other hand, the question of speed is completely irrelevant. If the environment in which a plant lives becomes cold, hot, or full of predators, the speed inherent in an animal's response has no meaning for it. What is much more important for the plant is to find an effective solution to the problem, something that will allow it to survive despite the heat, the cold, or the appearance of predators. To succeed in this difficult task, a decentralized, diffused structure is far preferable. As we shall see, this allows for more innovative responses and, being literally rooted, enables a much more refined understanding of the environment.

In order to come up with correct responses, it is essential to collect accurate data. It follows that plants, thanks to their choice to be rooted, have developed an exceptional sensitivity. Unable to escape from their environment, they manage to survive only because they can always and with great sophistication perceive a multiplicity of chemical and physical parameters, such as light, gravity, available minerals, moisture, temperature, mechanical stimuli, soil structure, gas composition of the atmosphere, and so on. In each case, the strength, direction, duration, intensity, and specific characteristics of the stimulus are all individually distinguished by the plant. Even biotic signals (from other living things), such as the proximity or remoteness of other plants, their identity and the presence of predators, symbionts, or pathogenic organisms, are stimuli, sometimes complex in nature, that the plant keeps recording and to which it always responds appropriately. This is further confirmation that the idea that plants lack sensitivity is nonsense.

Whereas animals react to changes in their surroundings by moving to avoid those changes, plants respond to the constantly changing environment by adapting to meet it.

SWARMS OF ROOTS AND SOCIAL INSECTS

There is a mystery still to be solved: How do plants manage without a brain, an organ that underlies every animal response? What systems do they use in its place? And, more generally, how do they manage to produce correct solutions to continuous environmental stimuli? The answer is a rather complex one, starting with the most important organ for rooted organisms: the roots themselves.

The root system is, without doubt, the most important part of the plant. It is a physical network whose apexes form a continuously advancing front; a front composed of innumerable tiny command centers, each of which supplements the information gathered during the development of the root and decides the direction of growth. Thus the entire root system guides the plant like a sort of collective brain or, better still, a distributed intelligence on a surface that can be huge. While it grows and develops, each root acquires information essential to the nutrition and survival of the plant. This advancing front can reach a really impressive size. A single rye plant is capable of developing hundreds of millions of root apexes. This is an extraordinary fact, yet negligible when compared to the root system of an adult tree. We do not have reliable data about the roots of trees, but certainly we are talking about several billion roots. We know that there can be more than a thousand root apexes in a single cubic centimeter of forest soil, but we do not have any realistic estimates of how many root apexes an adult tree might have in its natural environment. Today, the lack of techniques or tools that can record the movements of roots is the greatest obstacle to the progress of research into the behavior of plants. To obtain certain knowledge would require noninvasive and ongoing systems of three-dimensional image analysis of the entire root system; techniques that are still a long way off.

Despite the technical limitations, in recent years the study of roots has revealed unexpected aspects of how they work, for example the mechanisms they use to explore the soil. These processes have proven to be so efficient that they have been studied as a model for the construction of new robots. In the absence of predefined maps or points of orientation, exploration of unknown environments is not a simple task for instruments with a centralized organization. In contrast, a decentralized system, consisting of many small "agents," explorers operating

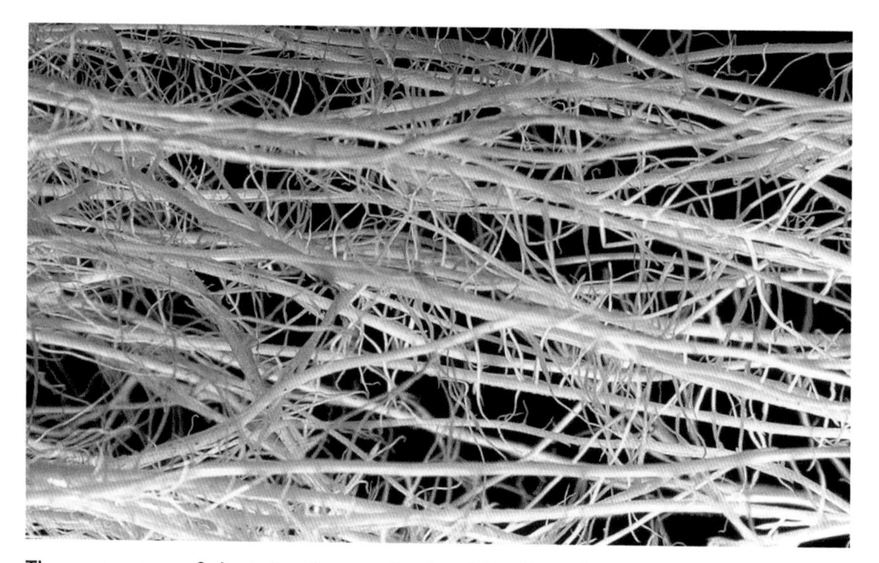

The root system of plants is a decentralized and distributed system, comprising millions of interacting units (root apexes).

in parallel, can probe the soil with far greater efficiency than a single robot, however sophisticated, ever could. As with the case above, we have relied more and more in recent years on the solutions found in nature to respond to technological problems, and not only in the plant world. A good example of organisms capable of collectively investigating unknown spaces—and therefore excellent models and sources of direct inspiration—is social insects.

Many animals, when acting as a group, display particular behavior. This is the case with swarms of insects or flocks of birds, which, through modalities of interaction, seem to act as a single organism. Similar collective behaviors have become an increasingly important field of research, not only to acquire basic knowledge about the functioning of the groups but also for the practical possibilities it opens up, enabling the application of the same systems in the most varied of technological solutions. The advantage that can be gained is twofold. First, these structures are particularly strong—not having an identifiable evaluation or communication center, they can withstand various types of stresses. And second, they have a simple design and are easily operated because they are based, even when developing apparently highly complex behaviors, on simple rules regarding the transmission of information between individual explorers.

For a long time, it was thought that these collective groups (swarms, flocks, schools, and so on) were formed only by animals. Yet, on a more abstract level, any group of single agents that makes autonomous decisions, lacks a centralized organization, uses simple rules to communicate, and, finally, acts collectively, is similar to such a community. And so it is with plants, whose modular structure can be equated to that of a colony of insects.

Thinking of the plant as a colony of modular parts is not a new

Flocks of birds are a classic example of the emergent properties of groups. All that is needed is simple rules to produce complex results.

idea. In ancient Greece, the philosopher and botanist Theophrastus (372–287 BC) wrote that "repetition is the essence of the plant," while in the eighteenth century, distinguished botanists such as Erasmus Darwin (1731–1802, a grandfather of Charles) and Johann Wolfgang von Goethe (1749–1832), the great scholar and botanist, thought that one should see trees as colonies of modules that repeat themselves. More recently, the French botanist Francis Hallé described the plant as a metameric (segmental) organism whose body consists of a set of coherent parts. Thus the recursiveness of the modules and the repetition of the hierarchical levels in the root system have enabled the study of roots using sophisticated mathematical methods typical of fractal analysis.

When observing the behavior of a root system engaged in the exploration of soil, we have found that even in the absence of a central nervous system, its growth model is far from chaotic. Indeed, it is perfectly coordinated and designed for the task it is to perform. For example, the roots have a surprising ability to detect very faint gradients of oxygen, water, temperature, and, in general, nutrients, and to follow these gradients to their source with great precision. However, how they are able to do that without becoming deviated by local variations, which are very common, has remained a mystery.

A few years ago, my colleague František Baluška decided to study roots as a collective organism, seeing them as being like a flock of birds or a colony of ants. This approach proved to be very effective, confirming that the structure of a plant's root system and the way it explores the terrain and uses resources can be described with great precision using swarm behavior patterns, such as those used in the study of social insects. Navigating along a tiny gradient is an almost impossible task for a single ant: any local variation in the gradient would cause it to get lost without the possibility of finding its way back. In contrast, by acting collectively, a colony can easily overcome this obstacle because it operates like a large integrated matrix of sensors that continuously processes the information received from the environment. We discovered that, like a colony of ants, root apexes work together, minimizing the inconvenience caused by local fluctuations.

As in a colony of insects, the protocol for transmitting information between one root apex and another, that is, between different autonomous agents, is highly likely based on "stigmergy." This term refers to a technique, typical of systems without a centralized control, that uses the changes in the environment as a means of communication. Typical examples of stigmergy have been observed in nature in ants and termites, which carry out wonderfully complex acts such as building nests with arches, pillars, rooms, and escape routes, starting with simple balls of mud. Stigmergy does not apply only to insects: even communication via the Internet, with messages left by users in a shared environment, is reminiscent of it.

Termites, like ants, produce colonies that are able to carry out extremely complex behaviors.

Plants are, therefore, organisms able to use resources that come from interactions between groups to respond to the problems and adopt very complex solutions. This capacity, which results from their diffused organization and lack of hierarchical levels, is so highly efficient that it exists almost everywhere in nature, including in numerous manifestations of human behavior.

A huge specimen of the *Ficus benghalensis* (or Banyan tree) with its gigantic aerial roots.

ATHENIANS, BEES, DEMOCRACY, AND PLANT MODULES

As many know, the term *democracy* comes from Greek (*krátos* of the *démos*, that is, dominion of the people) and describes in a rousing and precise way that wonderful transformation in the management of power that Athens gave to humanity around 500 BC and that since then has been the cornerstone on which our civilization is built. Perhaps less well known is that since then, the very concept of democracy, and thus the system by which the people manifest their power, has changed greatly, to the point that if an Athenian from the classical period awoke today in any "democratic" nation in the world, he would have great difficulty recognizing even the slightest affinity with the system of government to which he was accustomed.

The sovereign body of Athenian democracy consisted of the so-called assembly (*ecclesía*), composed of all citizens over eighteen. Its decisions, made by majority vote, had definitive value for legislative and governmental activities. Athenian democracy was a direct democracy, with a management of power that did not involve any intermediary, a huge difference from the system that we are used to today, which is more properly called representative democracy. Whether direct management of power is better or whether it is more efficient to delegate the burden of making the choices to representatives has been the subject of heated debate since antiquity. In his dialogue *Protagoras*, for example, Plato depicts a Socrates who is highly critical of the ability of the people, without adequate knowledge, to decide on issues of public life. He observes about Socrates:

> Now I observe that when we are met together in the assembly, and the matter in hand relates to building, the builders are summoned as advisers; when the question is one of shipbuilding, then the

> ship-wrights; and the like of other arts which they think capable of being taught and learned. And if some person offers to give them advice who is not supposed by them to have any skill in the art, even though he be good-looking, and rich, and noble, they will not listen to him, but laugh and hoot at him, until either he is clamoured down and retires of himself; or if he persist, he is dragged away or put out by the constables at the command of the prytanes. This is their way of behaving about professors of the arts. But when the question is an affair of state, then everybody is free to have a say—carpenter, tinker, cobbler, sailor, passenger; rich and poor, high and low—anyone who likes gets up, and no one reproaches him, as in the former case, with not having learned, and having no teacher, and yet giving advice; evidently because [the Athenians] are under the impression that this sort of knowledge cannot be taught.

The reasoning that Socrates used when disputing with the principle that the Athenian people should have the last word on everything that concerns the life of the polis resonated through all the criticism of direct management of power by the people, for how it has been embodied from the Athenian times of splendor until today. The fact that direct democracy defined perhaps the most fruitful period in the history of humankind is also considered a marginal detail by detractors of the system. Supporters of oligarchies (even contemporary ones) believe, by contrast, that the reasoning that they call "natural" is more interesting and effective: it is often argued that the formation of hierarchies—to put it crudely, survival of the fittest, the law of the jungle—is inherent in nature. We cannot escape such laws, however unpleasant. In *Gorgias*, another famous dialogue by Plato, Callicles says "the law is made by the weak for the weak. But nature itself shows that to be fair he who is worth more must prevail over he who is worth less, the capable over the incapable." In nature, hierarchies, intended as individuals or groups who decide for the community, are rare. We see them everywhere because we look at nature through human eyes.

Once again, our eyes see only what seems to be similar to us and ignore anything that is different from us.

Not only are oligarchies rare, but imagined hierarchies and the so-called law of the jungle is trite nonsense. What is more relevant is that similar structures do not work well. In nature, large, distributed organizations without control centers are always the most efficient. Recent advances in biology on the study of the behavior of groups indicate, beyond a doubt, that decisions made by large numbers of individuals are almost always better than those adopted by a few. In some cases, the ability of groups to solve complex problems is astounding. The idea that democracy is an institution against nature therefore remains just one of the more seductive lies invented by man to justify his (unnatural) thirst for individual power.

Animal communities have to continuously reach decisions about which direction to take, which activities to start, and how to carry them out. What are their behavioral models, in these cases? Are the decisions entrusted to one or a few, according to an arrangement that was described in an illuminating way by Larissa Conradt and Tim Roper (who have studied collective decision making in animals) as "despotic," or are they instead shared by the largest possible number of individuals in a "democratic" model? In the past, most scholars would have answered without hesitation: decisions in the animal world are the sole responsibility of one or a few members.

The banal reasoning upon which the confidence of their reply depended was the fact that democratic decision making is normally tied to two skills: voting and knowing how to count the votes, characteristics that are not so easy to find in nonhuman animals. Indeed, until recently and due to this insurmountable obstacle, any reasoning about the possible mechanisms of group decision making by nonhuman species was considered impossible. In recent years, however, the identification of particular movements of the body, sound emissions, positions in space, signal intensity, and a host of nonverbal means of communication has opened unimaginable perspectives on the capacity of animals to make group decisions.

In 2003 Conradt and Roper released a study on the methods by which animals implement shared choices. It is a clarifying study: the two authors emphasize that group decisions are the norm for the animal world, and they identify the "democratic" mechanism of participation as the most common method of making those decisions. Unlike the "despotic" method, the democratic method ensures lower costs for members of the community as a whole: even when the "despot" is the most skilled individual, if the group size is large enough, democratic practice ensures the best results. In short, widespread participation in making decisions is the system that evolution rewards most; group choices respond better to the needs of most members of the community even compared to dictation by an "enlightened boss." As Conradt and Roper wrote, "Democratic decisions are more beneficial primarily because they tend to produce less extreme decisions."

Bees illustrate the dynamics of behavior by a body of animals. Their predisposition to react in a social way is so pronounced that since antiquity—and long before expressions such as "swarm intelligence" and "collective intelligence" were dreamed up—it was clear to anyone who studied bees that their colonies are much more complex than the simple sum of the different individuals that compose them. In fact, bees show an organization that in its basic mechanism is reminiscent of the workings of the brain, with the individual playing the role of the neuron. This similitude occurs whenever the swarm must make decisions, such as in the formation of a daughter colony.

When a hive exceeds a certain size, it is necessary that the parent colony be split up to create a new one. So a queen bee, accompanied by some ten thousand workers, leaves in search of somewhere to found the new hive. Migrant bees fly away and travel, often a long distance,

The *Euphorbia dendroides* is a species typical of the Mediterranean maquis. It can create bushes with dichotomous stems and branches that can reach up to seven feet high.

from the parent hive, when they stop for a few days on a tree and do something surprising: some scouts scour the surroundings and return with information about the various possibilities. Then begins a real democratic debate, in the style of classical Athens. The bee group focuses on the question of how to choose the best place to set up the new colony among so many possibilities. Nature shows thousands of examples of such collective behavior; systems without a control center are everywhere. Even if we are not aware of it, even our individual decisions (those that belong to each of us) are made in a collective manner: the neurons in our brain, which produce thoughts and sensations, work the same way as the bees that must determine which is the best place for their new home. In both systems, the method of choosing is essentially a

Swarms of bees, like most animal groups, make shared decisions based on how much consensus has been expressed about the different options.

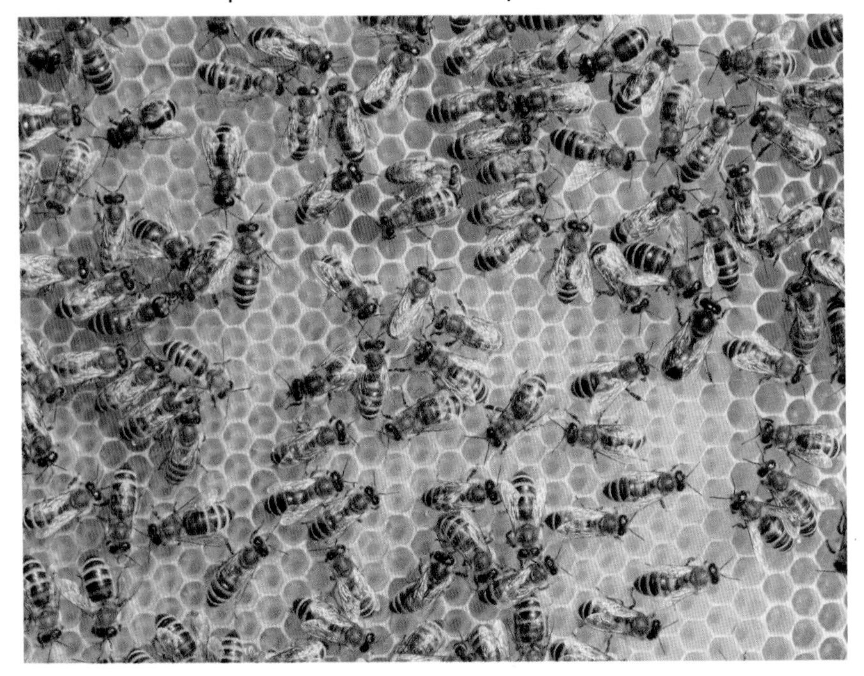

competition between different options: the one that receives the greatest consensus wins, whether determined by neurons that produce electrical signals or by insects that dance.

But back to our bees, whom we left dangling from a tree while scouts went around assessing the different options. Eventually the scouts return to report to the swarm about the features of the sites visited. The report is quite theatrical; it is a veritable dance, and the more the scout likes the place from which it has just returned, the more complex is the ballet. At this point other bees, attracted by the quality of the dance, go to visit the site in question and, upon their return, join in the propaganda ballet. The groups of dancing bees expand, the most publicized sites will also be the most viewed, and slowly the number of a site's supporters will increase. In the end, the location that has persuaded the largest group of bees will be chosen for the hive and the queen, with her swarm, will head in the direction of that site.

JURY THEOREM

There is a striking similarity between bees involved in deciding what is the most suitable place to found a new hive and the neurons in our brain that are occupied in considering alternative solutions to a problem. Like swarms of bees, the neurons in our brains individually have minimal information and minimal intelligence, yet collectively both bees and the neurons in our brains are capable of making correct decisions. These decisions are made through a real democratic vote between the members of the group: the largest number of bees that have visited a site or the largest number of neurons that have produced electrical signals dictates the final decision. This means that even our personal decisions are the result of a process of democratic choice, as happens everywhere in nature. The fact that where there are groups, similar systems develop attests to the existence of general principles of organization that make groups more intelligent than even the most intelligent individuals that compose them.

In 1785 Marie-Jean-Antoine-Nicolas de Caritat, Marquis de Condorcet (1743–1794), an influential French economist, mathematician, and revolutionary, developed a theory about the probability of a given group of individuals adopting a correct decision. This is the so-called jury theorem, which states that if the number of jurors increases, so do the chances that the group as a whole will reach the right decision. According to Condorcet, therefore, the effectiveness of a jury is directly proportional to the number of its members, at least if they are skilled and competent. In a group trying to resolve an issue, the chance of arriving at the best solution increases in proportion to the increase in its size. It would seem only a trivial mathematical transposition of the proverb "Two heads think better than one," yet Codorcet's theorem was the beginning of a revolution. It became the foundation of the processes of democratic decision making in relation to politics. In practice, however, his theorem proved to be much more, establishing the theoretical basis on which all subsequent studies on collective intelligence were founded. That same intelligence arises from the interaction of groups, which we have already seen at work in roots and insects and that underlies even the functioning of our brain. Any group of people, from families to businesses, from sports teams to armies, has experienced this result, and today, thanks to the sharing guaranteed by the Internet, humanity is becoming totally interconnected, exponentially increasing our collective global intelligence.

By collective intelligence, therefore, we mean the ability of groups to achieve results superior to those obtained through individual decisions, especially in solving complex problems—a principle with very promising application possibilities. Recently a team coordinated by Max Wolf of the Department of Biology and Ecology of Fishes at the Leibniz Institute of Berlin, published the results of detailed research on the capabilities of groups of specialized doctors in diagnosing breast

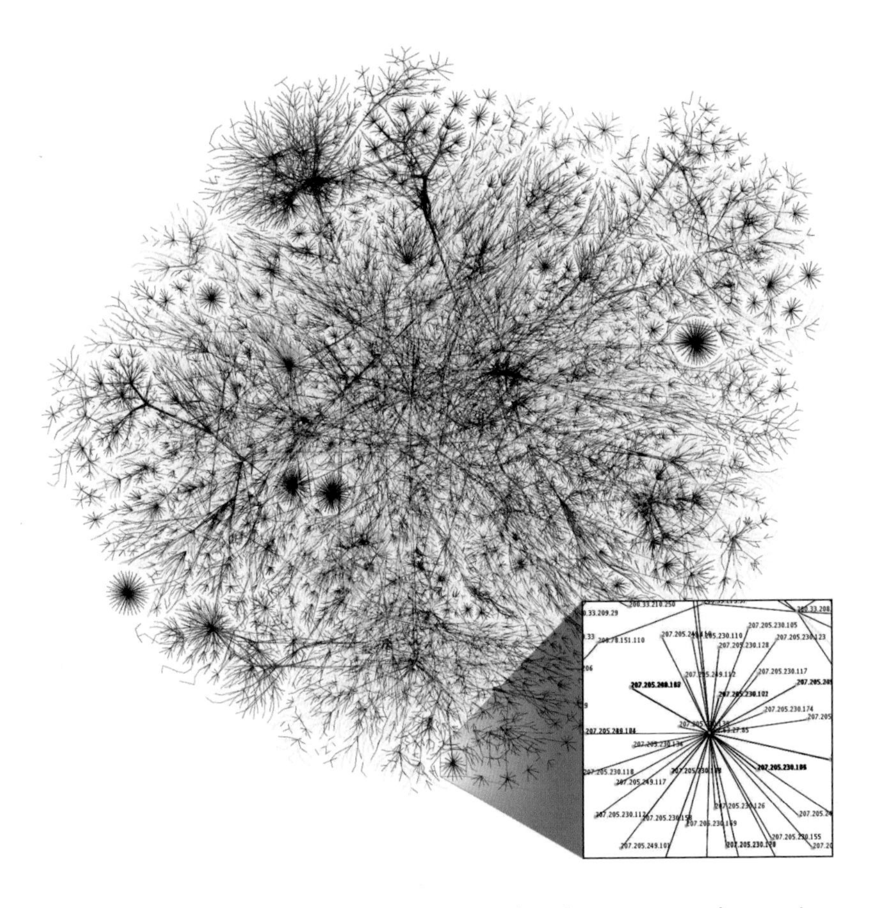

The topography of the Internet is very similar to that of a root system, because it responds to the same needs: a distributed system with no central command.

cancer with certainty on the basis of X-rays. It is a diagnosis that normally produces an average of 20 percent false positives and 20 percent false negatives. Wolf showed that the medical team, by using the usual instruments of collective intelligence, such as majority voting after meeting the quorum, achieved diagnostic results that were better than those of even the best individual doctors in the group.

Similar capacities have also been used more recently in solving scientific problems, with unexpected outcomes in different fields, such as determining the structure of a protein or the properties of nanomaterials.

In April 2016 a team of Danish physicists from Aarhus University showed that by using the capabilities of tens of thousands of players online, problems of quantum physics that have been unsolved for decades can be solved.

In the coming years, we will surely learn to harness the power of groups better and better. We are at the very beginning of a revolution that has much to teach us about the true nature of intelligence and that will involve increasingly large numbers of individuals in solving problems and achieving goals that today are impossible. The Internet is today's exemplar of the enormous power of nonhierarchical and diffused organizations that, like plant structures, is multiplying, gaining consensus, and, above all, producing excellent results. Wikipedia is an excellent example of how a colossal information resource can be structured by compiling the contributions of millions of collaborators. It has succeeded without any form of hierarchical organization or any financial incentive. It is an encyclopedia that in the English version alone had, by the end of 2017, 5.5 million articles, the equivalent of more than 2,000 printed volumes of the *Encyclopaedia Britannica*. If we consider the editions in various languages, Wikipedia has more than 38 million articles, equivalent to more than 15,000 volumes

How is it possible that without any hierarchical or administrative control, an organization can succeed? Wikipedia gives us a foretaste of what plant organizations are able to do, but it is only a start. The future I imagine will be ever richer. Models that renounce a vertical control of decision-making processes and in which all functions, even entrepreneurial ones, as well as any proprietary rights, will be ever more distributed. In fact, at least in Europe, similar structures—organized according to the plant model, distributed and rooted in the territory—have long existed; they are called cooperatives. They are organizations without any hierarchy that rely on the entire social structure; they are owned by their individual members, each of whom has one vote, and anyone can become a member. In terms of their structural characteristics, cooperatives are much more resistant to internal or external crises, and their failures are often due to having stopped acting like plant structures and instead turned into animal hierarchical organizations, thus losing flexibility and renouncing their knowledge of the territory.

Today examples such as cooperatives are critical to managing the transition to the new economy. Allowing this concept to correspond to the idea of web giants that accumulate huge profits in the hands of a few would be catastrophic. So, in addition to imitating the decentralized structure of plants in order to increase the creativity and strength of our organizations, we must imagine new forms of diffused ownership. In that sense, the tradition of cooperatives, combined with the extraordinary power of today's networks, can be a valid alternative model for the future. As with Wikipedia, it is hard to imagine what results could be achieved when the cooperative systems include the potential of the Internet and collective intelligence. Ancient Greece and Renaissance Italy were among the most creative moments in the history of Western civilization. In Greece, the city-states—geographically distant from one another—and forms of governance that often allowed every citizen to influence collective decisions gave rise to a period of unparalleled creativity in all fields of human knowledge. The same happened with the Italian city-states of the Renaissance.

In 2050 there will be ten billion people on earth, three and a half billion more than we are today. Many are alarmed by this enormous population growth, because they feel that we will not have enough resources to supply everyone. I do not belong to those ranks. Three and a half billion more thinking heads is not a problem but a huge resource. Three and a half billion people will be able to solve any problem—as long as they are free to think and innovate.

CHAPTER 5

THE CAPSICOPHAGOUS AND OTHER SLAVES OF PLANTS

When you are on the junk, the pusher is like the loved one to the lover.

—**WILLIAM S. BURROUGHS,** ***Junky***

Incidentally, the long-held idea that spices were used to mask rotting food doesn't stand up to much scrutiny. The only people who could afford most spices were the ones least likely to have bad meat, and anyway spices were too valuable to be used as a mask.

—**BILL BRYSON,** ***At Home: A Short History of Private Life***

PREVIOUS PAGE: The *Phelsuma ornata* is a small gecko endemic to Mauritius that acts as a pollinator for many plant species on the island.

THE ART OF MANIPULATION

Due to their obvious inability to move from the place in which they were born, plants often find themselves having to collaborate with animals, especially at particular moments of their lives. Plants can use animals' ability to move to disperse their seeds, to ensure efficient pollination, or for defense purposes. There are countless examples of such cooperation that have proven beneficial for both parties. Usually they offer a reward to the animal for services rendered. This is what happens when a pollinator is rewarded with tasty and energetic nectar, a bird gets a delicious fruit in exchange for spreading its seeds, or even humans—the best carriers on this planet—who in exchange for food, income, beauty, or other advantages spread the plants they need everywhere.

However, things are not always so clear. In many situations, the conduct of plants is more shifty and opportunistic, and the services provided by animals are used without any reward offered in return. The seeds of burdock—the plant that inspired the invention of Velcro—and hundreds of other so-called hitchhiker species cling to the fur of animals without offering anything in return for the lift. In many cases—and I am referring to the extraordinary mimetic capabilities of plants—these hitchhikers deceive the animals, forcing them into conduct that somehow facilitates or favors them. So far, nothing really new: deception, fraud, and misinformation are practices common to all living things, including plants. This whole subject becomes more interesting when we discover the true capacity for manipulation—and I use this word deliberately—that plants are capable of exercising on animals.

The wild teasel (*Dipsacus fullonum*) has long been the main tool with which to card wool, thanks to its structure, which evolved in order to cling to the fur of animals.

PUSHERS AND CONSUMERS OF EXTRAFLORAL NECTAR

In the mid-nineteenth century Federico Delpino (1833–1905), an important Italian botanist now sadly forgotten, and Charles Darwin corresponded extensively on extrafloral nectars. The subject interested them both, but they were advocates of diametrically opposed ideas. Many plant species are able to secrete nectar not only in flowers—the normal place for its production—but also on the branches, on the buds, or at the axils of the leaves. However, whereas the function of nectar in flowers is evident, thanks to its role as a lure and reward for pollinators, that of extrafloral nectar has long been shrouded in mystery. For Darwin, fluids emitted outside the flowers were to be regarded as waste products that the plant needed to eliminate. In other words, he argued, extrafloral nectaries were excretory organs used by the plant to expel substances that were in some way superfluous. He even held that, for future evolution, floral nectars originated from precisely those excretory organs.

Such a theory did not convince Delpino at all: that plants would waste such sugary, and therefore energetically expensive, substances seemed unlikely to him. A substance containing such a high amount of sugar could not be defined as superfluous. Delpino believed that if the plant deprived itself of such precious resources, it meant that in return

The *Nesocodon mauritianus* is a rare species endemic to the island of Mauritius. It became famous at the end of the 1980s, when it was discovered that it produced red nectar to attract the geckos that pollinated it.

it received benefits of some kind. His idea was that these substances had the same function as floral nectaries: to attract insects. There still remained the mystery of why a plant would draw insects to its body. In flowers the reason is obvious, but what useful function could insects flying between the branches and leaves have for a plant? The reason, discovered by Delpino after years of study, became known by the unattractive name of myrmecophily (from the Greek *múrmex*, meaning ant, and *phílos*, friend). In 1886 Delpino published a monograph listing three thousand myrmecophilous species—those that use extrafloral nectars to attract ants in exchange for an active defense system against other insects or predators in general. In essence, what he was describing was another of the many similar collaborations that plants create with animals: in this case, the sugary nectar in exchange for defense against predators.

Cooperation between plants and ants reaches levels of sophistication that are hard to imagine. An example is the association between these insects and numerous tree species belonging to the genus *Acacia*, native to Africa and Latin America. Some acacias produce special fruiting bodies to feed the ants and supply them with spaces, created inside specific structures of the tree, where the ants live and breed their larvae. But that is not all: as on one of those home shopping channels where the presenter never ceases to add products to entice you to buy, the acacias offer, in addition to food and shelter, free drinks in the form of the most welcome extrafloral nectar. In return, the ants are in charge of defense against any plant or animal that could somehow damage the plant on which they are hosted. And they do so very effectively. Not only do they keep far from the tree every other insect that has the unfortunate idea of approaching it, but they also attack with great vehemence even animals whose size is billions of times greater than theirs. So it is not uncommon to see ants bite herbivores the size of an elephant or a giraffe until they desist. The active defense implemented by the ants, however, is not limited to warding off animals, irrespective of their size; it goes far beyond that. Each plant that dares to come out of the ground within a radius of several feet from the host plant is mercilessly chopped off. So in the middle of the Amazon rain forest, it is not unusual to see perfectly circular areas devoid of any vegetation developing around an acacia—an unexplained phenomenon to local people, who call these areas "the Devil's gardens." In short, this arrangement between ants and plants seems to be a splendid form of collaboration for both parties; at first sight it is a classic example of mutualistic symbiosis. However, things are not exactly as they seem, and recently many studies have presented a more disturbing picture: under the guise of an idyllic, mutually beneficial relationship, there seems to hide, instead, a vile story of manipulation and deception, which sees the acacias in the unpopular role of the bad guy.

Ants find the extrafloral nectar produced on the branches, buds, or axil of the leaves of many species of plants irresistible.

The extrafloral nectar the plant produces, as we have seen, is a sugary liquid that is very energetic; everyone knows that nothing attracts insects more than sugar, so for years it was believed that this was the secret of the appeal of these secretions. However, the nectar does not contain just sugars; it also contains hundreds of other chemical compounds, including alkaloids and nonprotein amino acids. These substances play an important role of control on the animals' nervous system, regulating their neuronal excitability and therefore their behavior. One, for example, is the main inhibitory neurotransmitter in both vertebrates and invertebrates, including ants. Thus, alterations in ants' constitution due to the consumption of extrafloral nectar can significantly modify their behavior. In addition, the alkaloids contained in the nectar—including caffeine, nicotine, and many others—not only affect ants' cognitive capacity (as well as that of other pollinating insects that consume nectar) but they also create addictions. What has been discovered recently is that acacias, like many other species of myrmecophilous plants, are capable of modulating the production of these substances within the extrafloral nectar so as to modify ants' behavior. Not only that: like experienced dealers, acacias first attract the ants, luring them with the sweet nectar rich in alkaloids, and then, once the ants are addicted, they control their behavior, for example increasing their aggressiveness or their mobility on the plant—all this by modulating the amount and quality of the neuroactive substances present in the nectar. Not bad for beings that we continue to perceive as helpless and passive but that, precisely because they are rooted to the ground, have turned their ability to manipulate animals through chemistry into a true art form.

THE TIME I MET MY FIRST CAPSICOPHAGOUS

Do not think that we men are immune to the subtle spell with which plants are able to manipulate animals. On the contrary . . . Take chili peppers, for example.

I was born in Calabria, land of proud chili pepper eaters; there everyone, or almost everyone, loves spicy food. But not everyone is what I call a capsicophagous, or capsicum eater. These are a race apart, made up of people who have a unique relationship with the chili pepper. I met my first capsicophagous when I was a child, at a time when every encounter with original people, things, or facts was cloaked in magic and wonder. One of the most vivid memories from that time of my life is of a wedding my family attended. It was August, a month in which, if it were up to me, any kind of public or private ceremony should be formally prohibited. Putting on a suit and tie and attending the endless unfolding of a traditional southern Italian wedding—from the waiting in church to its finale on the dance floor—can result in more than fourteen hours of suffering, at temperatures far higher than those tolerated by any sensible person. Weddings in August are not something a responsible country should tolerate.

During that particular wedding, however, my attitude toward similar ceremonies changed once and for all. That was when I met my first capsicophagous, a real chili pepper eater. Let me be clear, I knew well the chilis used everywhere (my father, who converted in adulthood to the use of this spice, put large quantities in everything except coffee). It is impossible to find a Calabrese who does not more or less habitually eat spicy food, but the real capsicophagous is a different matter.

There were five of them, all in similar clothes, making them seem as if they were part of a fraternity: jacket, vest, and tie. Their dark clothes seemed very heavy to me as they approached the same table. Their

The term *chili* refers to the fruit of many species of the *Capsicum* genus (from the Solanaceae family) native to Mexico.

gestures were strangely synchronous, as though they were carrying out a tried-and-tested choreography. They moved the chairs at the same time, they sat down in unison, and then, moving at the same time as the others, each of them pulled out a big bunch of chili peppers from a pocket. As long as horns, all reds and greens. Beautiful. They lay them on the table with respect, next to their plate, between the glass of red wine and the fork—close to their left hand—and they settled to await the start of the meal. I was sitting at the next table, so I could see them clearly: serious people with little appetite for laughter and a slightly worried expression. They seemed to be waiting for something. They exchanged brief comments about the festivities and occasionally rested their hand on their chilies, almost tenderly, feeling the consistency of the individual fruits and, with sidelong glances, comparing them to those of the others. They had hands that were rough and darkened by the sun but that were capable of loving gestures toward those little spicy friends. The waiters began to serve, and finally I was able to observe the movements that all over the world reveal the precise body language of the capsicophagous. The right hand brings the food to the mouth while the left grips a chili pepper. That is how a capsicophagous tackles any meal: a morsel of food, a bite of pepper; morsel-bite, morsel-bite, without skipping a turn with any dish, precise like a metronome. It is the alternation of food-chili, on top of the impossibility of eating anything without accompanying it with something spicy, that characterizes the real capsicophagous.

I remember the scene perfectly: the rhythm of the action, remaining constant even in breaking off another pepper from the bunch when the previous one was finished. Everything worked like a perfect mechanism, oiled by years of practice. It was something I had never seen before that in my childish fantasy I tied to some exotic habit acquired by the five of them from distant lands. In later years I met many others in Calabria and everywhere else in the world, from China to Hungary, from Chile to Morocco and India, all sharing the need to accompany each dish with big bites of chili peppers. Regardless of the system used—forks, chopsticks, hands—at whatever latitude, the food-chili rhythm is the unmistakable hallmark of a staunch capsicophagous.

The fact that despite the terrible heat, the black and heavy clothes, and the enormous quantity of chili pepper consumed, those gentlemen did not sweat impressed me more than anything else. How was it possible? I was melting under the fiery Calabrese sun, while on their foreheads there was not the slightest drop of sweat to be seen. Dry as a bone. I was so curious that, after some time, overcoming my natural childish shyness, I risked asking one of them if the chilies they were eating with such determination were "normal" or of a different type, not spicy, and that they were maybe what was needed in order not to sweat. It was, as you can imagine, a careless move: never ask a capsicophagous if his chilies are spicy! His taste buds, cauterized by years of abuse, never feel enough burning. Hence the one I asked suggested with great kindness that I try a bit. A tiny bit. You know, just to get an idea. It was like putting lava in my mouth. A sensation we all know. Awful, that is all there is to say. Yet a little more than a third of the planet's population, about two and a half billion people, seek this very torment every day. How can that be?

The answer becomes clear when we examine the source of all the fuss. The name *chili* refers to a certain number of species of the genus *Capsicum*, the same that the sweet pepper belongs to, almost all of them producing a substantial quantity of capsaicin, the molecule responsible for the burning sensation (there are very few nonspicy varieties). The five most cultivated species are the *Capsicum annuum*, *C. frutescens*, *C. pubescens*, *C. baccatum*, and *C. chinense*. These are perennial shrubs that, however, given their short life, are usually treated as annuals. Originating in the Americas, where they were already being grown eight thousand years ago, these plants were very important to the native civilizations from a medicinal, as well as culinary, point of view. The chili pepper arrived in Europe thanks to Christopher Columbus,

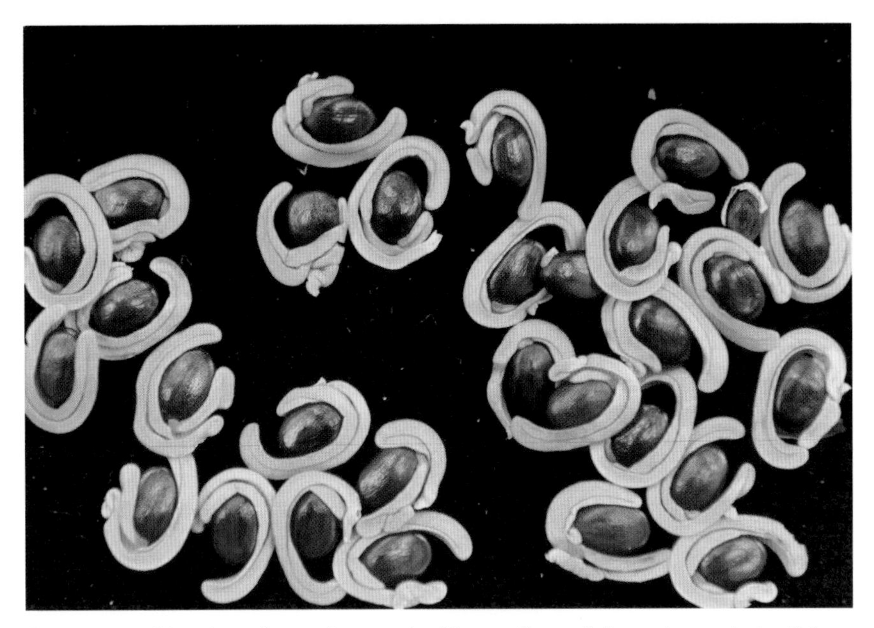

Acacia auriculiformis seeds are characterized by a yellow aril due to hyperplasia of the ovular funicle. This attracts ants thanks to its color and high nutritional value.

returning from his first trips to Central America, and like many other edible species native to the new world, chilis soon became a highly consumed plant, spreading globally. In less than a century the chili pepper became part of the gastronomic culture of countries and areas such as Italy, Hungary (where paprika comes from), India, China, West Africa, Korea, and so on. A unique and incomparable advancement in the conquest of the most remote places on earth, it is precisely the spicy quality that makes the chili pepper such a sought-after food.

To indicate its intensity, an American chemist named Wilbur Scoville invented a scale of spiciness, the Scoville scale, in 1912. The method used to measure it, the so-called Scoville Organoleptic Test, is to dilute the chili extract in a solution of water and sugar; a group of tasters continues to dilute the solution until they judge it completely devoid of spiciness. The number of dilutions corresponds to a value in Scoville heat units (or SHU): the higher they are, the spicier is the chili. A sweet pepper has zero SHU, while pure capsaicin has a value of 16 million SHU. The 16 million Scoville scale units represent the absolute maximum of spiciness for a chili pepper, a value that has the same allure as the great physical constants, such as the speed of light or the absolute zero of temperature. It is an insurmountable limit that is the Holy Grail for the capsicophagous.

Every year, using any technique—legitimate and illegitimate—to improve the plants, a huge number of extremely spicy new varieties or selections are produced. The goal is to always raise the limit, getting as close as possible to the unattainable perfection of 16 million SHU. The names of these new varieties leave nothing to the imagination. When naming new varieties of plants, we are accustomed to names that suggest grace, sincerity, friendship, and beauty; the names given to these monsters of spiciness are notable for their brutality. Hell, devil, nuclear, death, ghosts, and plagues are among the most commonly used terms, but also tigers, scorpions, vipers, cobras, Komodo dragons, tarantulas, or similar animals are well represented in the imagination of the creators of new chilies.

In 2013 the Carolina Reaper (a monster capable of producing fruits containing more than 10 percent of their weight in capsaicin) exceeded the astronomical figure of two million Scoville units, thus ousting the Trinidad Scorpion and the Naga Viper from the coveted record of being the spiciest capsicum on the planet. Each year the bar rises, new world records of spiciness are achieved, and millions of people around the world search desperately for samples, either to taste them or to propagate them. And the spicier they are, the more they are propagated. Because what the capsicophagous is seeking is the capsaicin and nothing else, in increasing doses. In the United States there is even a spicy sauce for sale (and "spicy" is a euphemism) called 16 Million Reserve: this is made of pure capsaicin crystals, preserved in small bottles produced in

SCOVILLE PEPPER SCALE

TYPES OF PEPPERS	SCOVILLE HEAT UNIT
TRINIDAD SCORPION	1,463,700
BHUT JOLOKIA (GHOST PEPPER)	1,041,427
RED SAVINA HABANERO	250,000 - 577,000
CHOCOLATE HABANERO	200,000 - 385,000
SCOTCH BONNET	150,000 - 325,000
ORANGE HABANERO	150,000 - 325,000
FATALI	125,000 - 325,000
DEVIL TOUNG	125,000 - 325,000
KUMATAKA	125,000 - 150,000
DATIL	100,000 - 300,000
BIRDS EYE	100,000 - 225,000
JAMAICAN HOT	100,000 - 200,000
BOHEMIAN	95,000 - 115,000
TABICHE	85,000 - 115,000
TEPIN	80,000 - 240,000
HAIMEN	70,000 - 80,000
CHILTEPIN	60,000 - 85,000
THAI	50,000 - 100,000
YATSUFUSA	50,000 - 75,000
PEQUIN	40,000 - 58,000
SUPER CHILE	40,000 - 50,000
SANTAKA	40,000 - 50,000
CAYENNE	30,000 - 50,000
TOBASCO	30,000 - 50,000
AJI	30,000 - 50,000
JALORO	30,000 - 50,000
DE ARBOL	15,000 - 30,000
MANZANO	12,000 - 30,000
HIDALGO	6,000 - 10,000
PUYA	5,000 - 10,000
HOT WAX	5,000 - 10,000
CHIPOTLE	5,000 - 8,000
JALAPEÑO	2,500 - 8,000
GUAJILLO	2,500 - 5,000
MIRASOL	2,500 - 5,000
ROCOTILLO	1,500 - 2,500
PASILLA	1,000 - 2,000
MULATO	1,000 - 2,000
ANCHO	1,000 - 2,000
POBLANO	1,000 - 2,000
ESPANOLA	1,000 - 2,000
PULLA	700 - 3,000
CORONADO	700 - 1,000
NUMEX BIG JIM	500 - 2,500
SANGRIA	500 - 2,500
ANAHEIM	500 - 2,500
SANTE FE GRANDE	500 - 750
EL-PASO	500 - 700
PEPERONICINI	100 - 500
CHERRY	0 - 500
PIMENTO	0
BELL PEPPER	0

Source: The Scoville Food Institute — http://www.scufoods.com/REVtshirtart0811.jpg

BUSINESS INSIDER

The Scoville scale measures the spiciness of chilies. It goes from the zero of the sweet pepper to the approximately 1,500,000 Scoville degrees of the spiciest varieties.

The Carolina Reaper is the current world record holder for spiciness. Every year over 30 million tons of chilies are produced, using an area of approximately 25 million acres.

limited quantities; its price on the market can easily reach thousands of dollars.

But what, exactly, is capsaicin? It is an alkaloid that, coming into contact with the nerve endings, activates a receptor known as TRPV1. This receptor has the task of reporting to our brain levels of heat that are potentially harmful, and indeed it is normally activated at around 109°F.

The TRPV1, in practice, has been "designed" to prevent us from doing dangerous things such as picking up a red-hot iron with our bare hands or putting boiling soup into our mouths, actions that could harm our body. That is why capsaicin causes pain, and that is why it is used by half the police forces around the globe as a weapon in the form of pepper spray. For the very same characteristics, however, it is appreciated as a spice. Yet sane people do not put lemon juice into their eye or slam their shins against furniture because they find the pain that ensues pleasurable.

So why does a third of the world's population love to put on their tongue—one of our most sensitive organs—large quantities of an alkaloid that causes a terrible burning sensation? Several theories about it have been developed over the years. The best known, cited whenever you try to explain this abnormal human behavior, is what the psychologist Paul Rozin has termed "benign masochism," according to which a certain type of person is drawn to sensations of danger, including burning. For these people, eating hot peppers is a variant of going on a roller-coaster ride: in both cases, Rozin maintains, despite the fact that the body perceives the danger of such an activity, at a deeper level it knows that there is no real danger, and so there is no real need to stop the negative stimulus. Rozin concluded that, after a series of exposures to the same stimulus, the initial discomfort becomes pleasure.

Although I appreciate the elegance of it, such a theory has never convinced me because I relish the spicy food but riding roller coasters, bungee jumping, and other similar activities do not attract me at all; because my wife, who loves spicy food, covers her face to avoid seeing gory scenes in horror movies and would not even go on a swing, let alone on a roller coaster; because many of the voracious capsicophagi that I have known are among the most mild people I have ever met and the least likely to seek dangerous sensations; and finally, because it seems unlikely that a third of the world's population meets these criteria, which, at a rough guess, do not seem to be so widespread. But I could be wrong. For it should be said in Rozin's favor that research carried out in 2013 on ninety-seven subjects by two food science scholars, John Hayes and Nadia Byrnes, established a significant correlation between those who "sought thrills" and those who liked spicy food.

My hypothesis, rather, is that so many people in the world love the spiciness of the chili pepper because capsaicin provokes a different effect from that induced by other plant alkaloids that act directly on the brain (such as caffeine, nicotine, morphine, and so on) but is identical in its ultimate aim: to induce addictiveness. When the body senses pain on the tongue, it gives the green light to a cascade of signals to the brain, which produces endorphins to relieve the suffering. Endorphins are a group of neurotransmitters with analgesic and physiological properties similar to those of morphine but much more powerful. They are the system by which our body relieves pain and are the key to understanding the hidden power that the chili pepper exerts on our lives.

Endorphin addiction is not a bizarre concept, far from it; for example, it is the mechanism behind the well-known runner's high. If you are a fan of running or have friends who engage in endurance sports such as marathons, open-water swimming, and cycling, you may have heard them talk about it: it is a special state of euphoria that develops after a prolonged and strenuous sporting performance. Comparable to the high induced by certain drugs, it may manifest itself as intense happiness or a strong feeling of well-being. For many years there has been no scientific evidence that the phenomenon was real. Rather, it was seen as a myth of running enthusiasts, at least until 2008, when a survey carried out in Germany on athletes who were analyzed before and after an intense sporting activity proved that it was well founded. The runner's high is a real phenomenon that manifests itself through the release of endorphins in the brain. The analgesic potency of this substance also explains the high threshold of pain that can often be seen in athletes undertaking intense physical activity. There are numerous cases of marathoners who have kept running despite fractures or injuries that in other circumstances would have been unbearable. It is the same mechanism by which those who consume large quantities of chilies tend to be less sensitive to the pain; the anesthetic capacity of the capsaicin has, in fact, been well demonstrated in the scientific literature of recent years.

Just like many other plants that produce addictive substances, the chili turned to chemistry to secure the most powerful and versatile mammalian carrier: humans. I believe that what makes this plant even more interesting is that, unlike other plant-based drugs that also exert activity on the brains of other animals, the capsaicin affects only humans. There are no cases of other mammals that like eating the pepper fruit.

It seems that at the beginning of its evolutionary history, capsaicin had the ability to induce a certain resistance to fungal infection in the plant. Thus, in areas with the greatest number of such attacks, the capsicum fruit started to naturally contain a higher concentration of this alkaloid. Later, a further evolutionary advantage was the lucky circumstance that birds lacked the receptor charged with causing the burning sensation in mammals, facilitating the seed dispersal of the spiciest plants. In fact, the capsaicin keeps mammals away because their chewing destroys the seeds contained in the fruit, while birds, which are much more reliable carriers, do not even notice the capsaicin because they do not chew the seeds, and therefore they transport them further afield. But the real advantage of capsaicin for the chili pepper has been its ability to bind itself through atypical dependency to humans—the ultimate vector.

If my theory about the conditions of slavery to which the alkaloid

Turtles from the Galápagos islands and Mauritius were once among the most important transporters and diffusers of seeds on their respective islands.

has reduced us capsicophagus mammals still does not convince you, you will have to visit one of the thousands of chili fairs that are held every year in any country in the world. The environment in which the capsicophagous of the third millennium operates is different from the traditional and darkly clothed one of my childhood. At the fairs, you can meet new followers studying the composition of sauces with names from horror or apocalyptic movies while wearing hats sporting a drawing of a capsaicin molecule—the most fanatical tattoo the structural formula on their throats—and T-shirts with the words Pain Is Good. The use of chili pepper is constantly increasing in the world. Populations traditionally immune to the insidious pleasure of spicy cuisine are consuming it in quantities and ways that were unthinkable a few years ago. In short, the strategy that this species has implemented to make humans dependent on it, and to put them at its complete service, has proven to be successful. Associating itself with humans has allowed it to spread around the entire planet in a few centuries; no other carrier could have done anything like that in such a short time. And in the future things will get better and better: ultimately, to get an endorphin rush it is simpler, and a lot less work, to turn to a nice plate of chili peppers instead of doing a race of 26.219 miles.

CHEMICAL MANIPULATION

The example of the chili pepper and its alkaloid is hardly an isolated case. Many chemical compounds of plant origin affect the functioning of the brain, and the mechanisms through which these psychoactive molecules operate are well explained. What is not clear is why compounds produced by plants have an effect on the brains of animals. Indeed, why should plants spend energy on producing molecules that cause effects on the brains of animals?

The current neurobiological theories about drug use are based on the observation that all molecules that are addictive activate an area of the brain involved in managing the reward system. Every time we do something useful for our survival, this ancient area of the brain—which evolved to activate in response to stimuli, such as food, water, and sex—rewards us with pleasure, thus causing us to repeat the action. Acting on the same system, the drugs induce us to repeat the consumption of the molecule that triggered the reward system, thus creating dependency.

Yet all the hypotheses on the origin of plant-based drugs consider the main alkaloids (caffeine, nicotine, and so on) as neurotoxins that were developed to punish and deter herbivores. According to this theory, evolution should not have produced compounds that, acting on the reward

The lotus has been one of the most famous and admired aquatic plants since ancient times. Its impressive flowers in white, pink, red, and blue hold a strong appeal for pollinating insects.

system, increase the consumption of the plant. In an ecological context, this apparent contradiction is known as the "drug-reward paradox." However, if we accept the idea that the neuroactive molecules produced by plants are not a mere deterrent but rather a tool with which to attract animals and manipulate their behavior, the paradox is easily resolved by placing the plant-animal interaction in a very different ecological context and opening new angles in the neurobiological search for effective tools to combat drug abuse. Going back, therefore, to the extrafloral nectar at the beginning of the chapter, the relationship between plants and ants and their long history of coevolution offers the ideal model on which to test this hypothesis. And if, as I believe, we can show that even in the interaction with ants, the production of neuroactive molecules is used by plants to manipulate their behavior, we will have further evidence of this not inconsiderable ability, which would radically change our image of plants from simple beings, passively at the mercy of animal needs, to complex living organisms capable of manipulating the behaviors of others—a notable reversal of roles.

CHAPTER 6

ARCHIPLANTS

The materials of city planning are the sun, trees, sky, steel, cement, in that hierarchical and indissoluble order.

—**LE CORBUSIER**

A doctor can bury his mistakes, but an architect can only advise his clients to plant a Virginia creeper.

—**FRANK LLOYD WRIGHT**

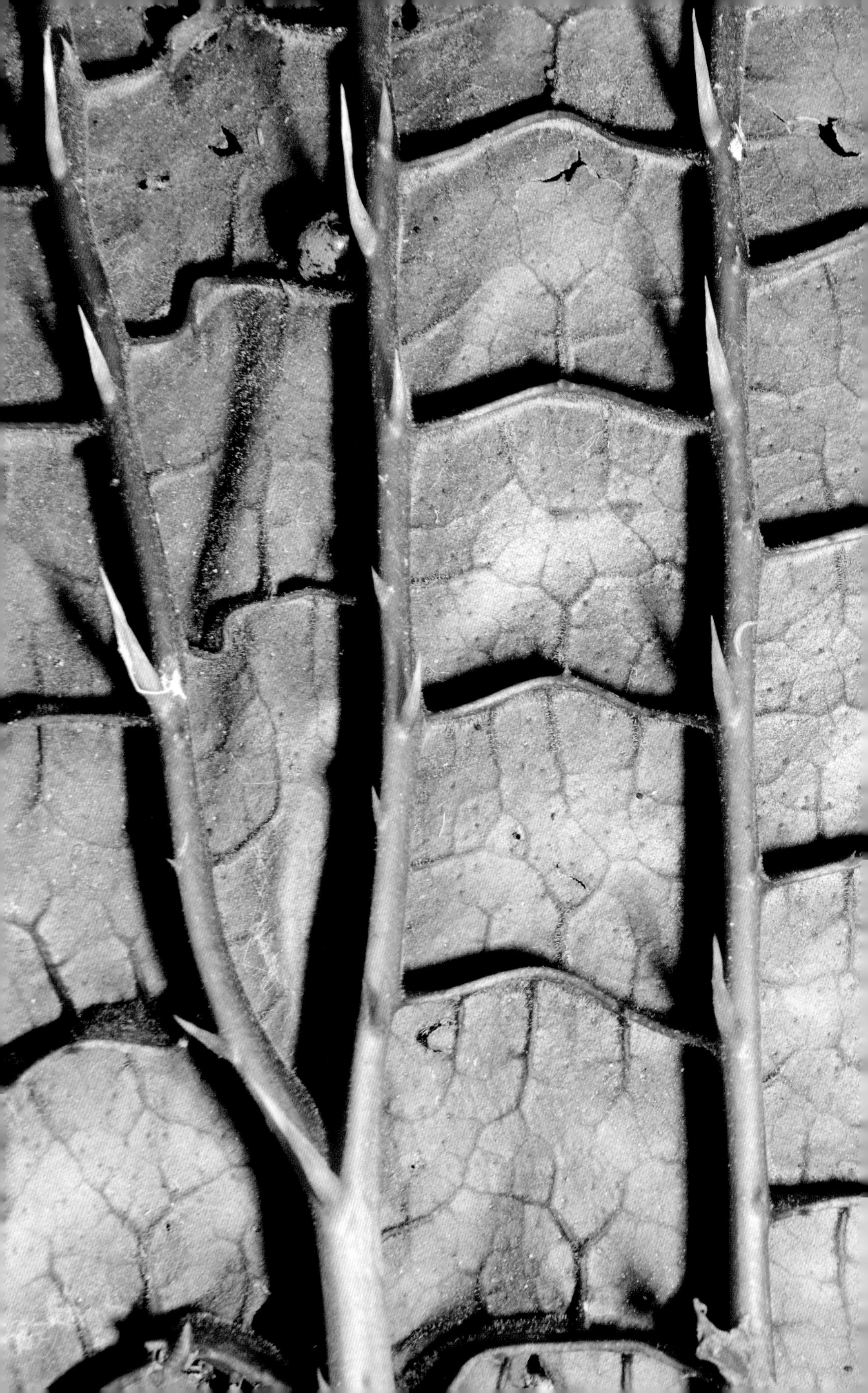

PREVIOUS PAGE: An image of the nervation of the underside of a *Victoria amazonica* leaf. Thanks to this structure the leaf can bear enormous weight.

TOWERS LIKE BRANCHES

Among the many talents of Leonardo da Vinci, his great powers of botanic observation are among the least recognized. To Leonardo we owe thanks for some fundamental discoveries about the nature of plants, such as the explanation of how the annual rings of secondary growth on the trunk of a tree are formed and how, from the study of the number, distribution, and thickness of those rings, we can pinpoint the age of the tree and the climate during every year of its life. It was Leonardo's intuition that the growth resulting in the widening of the trunk is a consequence of the action of a specific tissue, only much later identified as the so-called cambium: the growth in the diameter of trees is produced by the sap, which, in the case of trees, is generated in the month of April between the outside coating (camisia) and the wood of the tree. At the same time this outside coating converts into bark.

The discovery that concerns us here, however, relates to the principle according to which the leaves are arranged on a branch, so-called phyllotaxis (from the Greek *phýllon*, leaf, and *tàksis*, arrangement). Leonardo covered the basic concepts with extreme accuracy, centuries before the Swiss botanist Charles Bonnet (1720–1793), who is commonly recognized as its discoverer. What exactly is phyllotaxis? If you look closely at the succession of leaves on a single branch of different plants, you will see that each one follows a particular rule: some are arranged on a more or less narrow spiral, others perpendicularly on consecutive layers. Every species has its own rule of succession in the arrangement of its leaves. This would not seem, at first glance, to be something that is very interesting or that could have any practical application beyond the

Phyllotaxis is the arrangement of leaves along a stem. Each plant has its own phyllotactic formula.

taxonomic classification of plants; nor does it appear to be a discovery that would somehow affect how we construct our buildings. Yet it is all of those things.

Leonardo is certainly not just any old scientist. Not content to describe a phenomenon, he also wanted to understand the causes that generate it and, if possible, turn his findings into practical applications. Therefore, he provided a functional explanation of phyllotaxis: it is the arrangement that guarantees the leaves the best light exposure, without their overshadowing each other. The result of hundreds of millions of years of evolution, this arrangement can be copied and put to good use. That is what the Iranian architect Saleh Masoumi did in his surprising plan for a phyllotactic tower. Inspired by the way leaves are arranged along a branch, he designed a residential skyscraper that has some unique features.

One of the problems common to apartments in any building or residential tower is that normally they are surrounded by other dwellings, with no direct access to the surrounding environment. Usually, the ceiling of the apartment below is the floor of the apartment above. Under these conditions, of course, the amount of light received by each unit is only a fraction of what is possible. Masoumi's tower solves this problem rather brilliantly: by organizing the apartments into a phyllotactic arrangement around the central axis of the building, each one receives light from all sides, like a leaf on a branch. Each unit even has access to the sky above, with the possibility of collecting sunlight to be used for energy purposes.

In fact, there is no better model than the phyllotactic one to expose surfaces at different levels to sunlight. Evolution, through a long process of trial and error, has selected only the results that ensure an optimal uptake of light by individual leaves. That same solution, if applied to buildings, could guarantee energy results that are unimaginable and revolutionize the way we think about the structure of our buildings. Maybe Leonardo, in his genius, already predicted that one day, thanks to the study of the arrangement of leaves, new towers would be designed. This is yet another fascinating example of how science, whatever the object of its interest—plants included—produces results whose applications are often entirely unpredictable. In fact, much of science's charm lies within this very unpredictability.

THE *VICTORIA AMAZONICA*: HOW A LEAF SAVED THE FIRST GREAT EXHIBITION

The epic of the *Victoria amazonica* begins in the first half of the nineteenth century, a plant destined to become a case study not only for botany but also for architecture. The story of this plant is very Byzantine, even in the attribution of its name. Its seeds and a description of it arrived in France in 1825, shipped from South America by the French naturalist, botanist, and explorer Aimé Bonpland (1773–1858), who, however did not disclose his discovery and did not give a name to the new species. In 1832 the German explorer Eduard Friedrich Poeppig (1798–1868) reported on the plant in the Amazon and published the first description, calling it the *Euryale amazonica*. Finally, in 1837 the English botanist and pioneer of plant classification John Lindley (1799–1865) renamed it in honor of Queen Victoria, thus kick-starting its botanic fame. We are interested in this species because, in addition to fascinating audiences around the world for its elegance and size, it sparked the imagination of architects and engineers due to the extraordinary strength of its enormous leaves. The *Victoria amazonica*—today one of the superstars of any self-respecting botanical garden—quickly became celebrated well beyond a small number of scholars and enthusiasts and established itself as a popular icon of the late nineteenth century. Its images were printed on fabrics, books, and wallpapers; wax reproductions of its flowers became all the rage, and illustrations with children kept afloat without difficulty by the enormous leaves stimulated curiosity about this exotic aquatic plant. Obviously, its extraordinary structural capabilities did not escape the attention of experts: How could a single leaf withstand a load, if properly distributed, of up to nearly a hundred pounds without breaking or deforming? But above all: Would it be possible to replicate this marvelous feature in man-made designs?

The *Victoria amazonica* has leaves that look like large circular trays and can measure up to more than eight feet across, with raised edges that are anchored to the bottom of the calm waters where they live by long steles, which arise from an underground stem buried in the mud. The upper side of the leaf is waxy and when wet is covered with drops that slip away; the underside, which is fuchsia colored, is equipped

The leaves of the *Victoria amazonica*, an aquatic plant of the Nymphaeaceae family, can reach up to two and a half feet in diameter.

with spines that serve to protect it from fish and manatees that feed on aquatic plants. The air trapped in the spaces between the nervations enables the leaves to float. Each plant produces between forty and fifty leaves, which cover the surface of the water, thus blocking the light and limiting the growth of most other plants.

In 1848 the *Victoria amazonica* crossed paths with Joseph Paxton (1803–1865), the head gardener of William Cavendish, the sixth Duke of Devonshire, at Chatsworth House in England. Thanks to his undisputed skill in plant cultivation, from an early age—he was just twenty-three—Paxton was hired at Chatsworth in 1826 with the task of looking after the Duke's gardens. As often seems to be the case with members of the British aristocracy, Cavendish had a true obsession with plants: he owned a private botanical garden that was one of the most important in the world, with large greenhouses and dense arboretums.

Even bananas were grown at Chatsworth House. It was a banana from Mauritius that Joseph Paxton, with his legendary ability, was able to propagate at Chatsworth House, dedicating it to his patron under the name of *Musa cavendishii* (today more than 40 percent of the bananas present on our tables are Cavendish, derived from that species).

Another national characteristic of the British is their passion for competition. So it was that William Cavendish and the Duke of Northumberland engaged in a fierce challenge over who would first succeed

Joseph Paxton photographed with a fine specimen of the *Victoria amazonica* leaf inside a greenhouse at Chatsworth House.

The leaves of the *Victoria amazonica* have a structure that enables them to support a surprising amount of weight.

in the growing and flowering of the *Victoria amazonica*. The Duke of Devonshire relied upon Paxton to win, and the choice proved to be perfect. Thus in 1848 the head gardener received a seed from Kew Royal Botanical Gardens, a seed that within a few months he induced to bloom thanks to meticulous attention in reproducing the climatic conditions of its original habitat within a heated greenhouse. The flowers of this plant with enormous leaves became one of the major attractions of Chatsworth House, and Queen Victoria herself—to whom Paxton, with great foresight, had made a gift of a magnificent specimen—visited, accompanied by the French president, Napoleon III (who later would become emperor).

The uniqueness of this flower can be seen from the process used to ensure its own pollination. *Victoria amazonica* flowers have a relatively short life—around two days—and are initially white. On their first night, when they open, a sweet smell similar to that of pineapple attracts the beetles that are in charge of the transport of pollen. In addition, the plant, to make sure the pollinators come in large numbers, raises the temperature of the flower through a thermochemical reaction. This is a skill called endothermy or thermogenesis, possessed by a very small number of plant species (only 11 of about 450 known families of flow-

The structure of the underside of the *Victoria amazonica* leaf, used by Joseph Paxton for the design of the Crystal Palace.

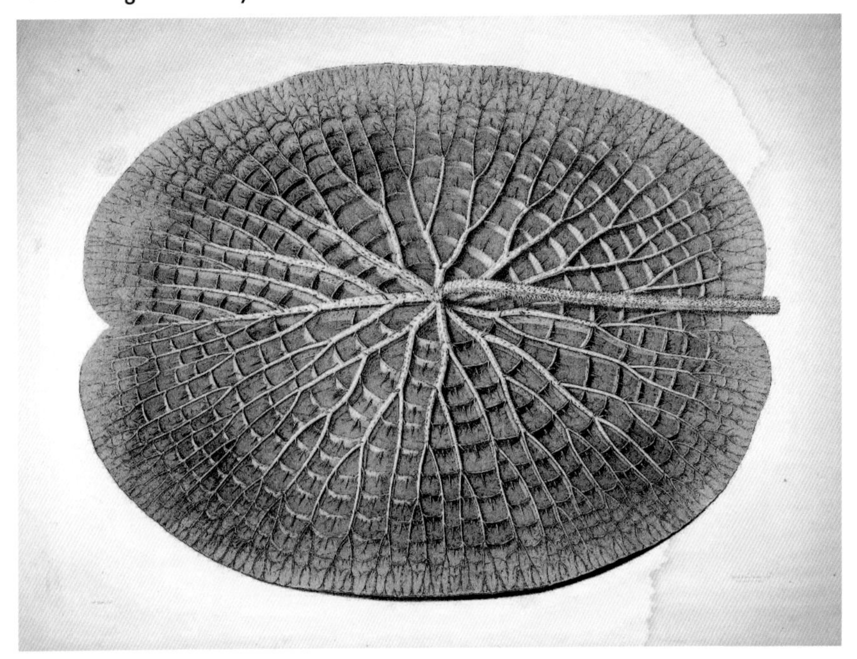

ering plants show thermogenetic properties), and in all these plants, the production of heat is linked to attracting pollinators. There are several reasons for this: heat can be a direct reward for the insect, it may increase the volatilization of chemicals that act as a draw, it can mimic the temperature of the feces of mammals, or it can stimulate oviposition (egg laying) in flies. In any case, if a flower produces heat, it does so to attract its pollinators, and the *Victoria amazonica* is no exception to the rule.

At this stage, the flower is exclusively female and is ready to receive the pollen collected by the beetles from other plants. Having penetrated inside, the insects transfer pollen onto the stigmas, enabling fertilization. Meanwhile, the petals close, blocking the insects inside until the fol-

The flowers of the *Victoria amazonica* are white the first night after flowering and become pink the next night. Their pollination is carried out by beetles.

lowing evening. The morning after, the flower changes, assuming male characteristics: the anthers mature and produce pollen. When it reopens in the evening, it has even changed color; it has become purple to indicate successful fertilization, it does not emit any smell and is no longer warm. The insects are free to go out, completely covered with pollen, and to repeat the entire process on another plant (each one has only one white flower at a time). After being fertilized and having released the pollinators, the flower closes and immerses itself below the water level.

In 1848 this process was entirely unknown. At that time, just getting the flower to bloom was an achievement worthy of the king of gardeners. And in fact, Paxton's reputation went beyond the confines of plant enthusiasts, extending even to nonexperts. It was only the beginning of a compelling story; the successes that the *Victoria amazonica* brought him were far from over.

In London in 1851 the first Great Exhibition was being planned. It would be a one-off event, and to host it properly, a colossal structure had to be built inside Hyde Park, one suitable for welcoming delegations from around the world as well as millions of expected visitors. The history-making occasion demanded a magnificence reflecting the greatness of the British Empire, and so the requisites for the design of the exhibition building were many. First of all, the structure was not meant to be permanent, and it had to be built quickly.

The cost was another consideration: in deference to the principles of sobriety that had made the empire great, the selected structure would have to meet the demands of functionality with minimal expense. Students of architecture across Europe took part in the competition for its design. The commission received 245 proposals, and, after a long process of analysis . . . it rejected them all.

The evaluation, however, had taken a lot of time, and no one had imagined that not one of the many designs submitted would prove suitable. There were now just a few months to go, and still no one had any idea how to host the Great Exhibition. In Parliament, newspapers,

Panoramic view of the Crystal Palace built inside Hyde Park in 1851 to house the first Great Exhibition.

and pubs there was talk of nothing else: How to respond so quickly to such a great challenge? Four experts were appointed to see to the design and the construction of the pavilion in little time. Unfortunately, even that solution failed. Great Britain ran the risk of making a sensationally bad impression before world public opinion. The exhibition that was supposed to celebrate the technological innovations and resourcefulness of the empire was in danger of turning into a fiasco. In this atmosphere of last resort, Joseph Paxton came forward with his revolutionary idea: to build a huge glass and cast-iron structure using prefabricated units. It was a brilliant idea that made history. Paxton presented a design of colossal dimensions. It was a building of 990,000 square feet: 1,851 feet long, 407 feet wide, and 128 feet high, so large that it could hold four Saint Peter's Basilicas inside it. It would have been impossible to achieve a construction of such proportions, within the permitted time and costs, without the tremendous insight of using identical prefabricated units. In those years British technology had evolved sufficiently to make the rapid production of the tens of thousands of units needed possible. The basic unit consisted of a square with sides measuring about twenty-five feet. Thus, by gradually adding new elements, the initial structure could extend to infinity. Even the exhibition spaces were calculated based on the number of units needed for each.

Mass production would require much less time and the costs would be infinitely lower than those of any conventional masonry building. Furthermore, if the whole thing were to be dismantled at the end of the exhibition, the various parts could easily be put to another use. Basically, Paxton proposed to erect a colossal glazed greenhouse so big that the trees of Hyde Park in the area concerned would be incorporated within it. He had already designed similar structures to protect the valuable exotic plants in the Cavendish collection against the cold British weather. Among those heated greenhouses the most impressive was the Great Stove, a huge tropical nursery, so vast that it could be visited in a carriage. Of course, that was nothing compared to what Paxton achieved for the Great Exhibition.

A building of this size, however, had to comply with strict structural requirements. In addition, the work had to be completed on schedule and at low cost. That is where Paxton's second ingenious intuition came in: to mimic the nervation of the *Victoria amazonica* leaves to create large arched vaults. Both bioinspirations—the use of units for the construction of the huge building and the use of the radial structure of the *Victoria amazonica*—were the result of the extraordinary passion the man had for botany.

More than two thousand workmen labored industriously at making what, according to a quip in the famous satirical magazine *Punch,* began to be known as the Crystal Palace. In just four months, the building was completed. Thanks to Paxton and the *Victoria amazonica*, London was ready to host the first Great Exhibition, with the pomp and grandeur appropriate to an imperial power.

The Crystal Palace left more than fourteen thousand exhibitors

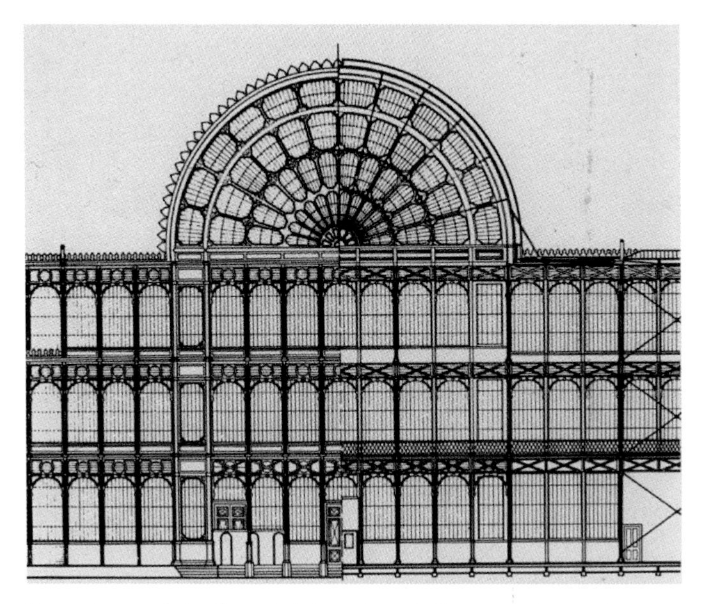

Frontal view of the Crystal Palace. Joseph Paxton designed the radial ribs of the barrel vault based on the structure of *Victoria amazonica* leaves.

from all over the world speechless and soon became the calling card for the technological innovation capacity of the British Empire. The event was a memorable success. More than five million people (a quarter of the British population at the time) visited the exhibition, among them Charles Darwin, Charles Dickens, Charlotte Brontë, Lewis Carroll, George Eliot, and Alfred Tennyson. The proceeds of ticket sales, net of costs, were used for the construction of the Victoria and Albert Museum, the Science Museum, and the Natural History Museum, as well as to set up a fund for scholarships in industrial research that is still active today. Paxton, the hero who made that prodigious work possible, was made a baronet. He never forgot the *Victoria amazonica* and botany, which remained his great passions and began a business career that made him quite wealthy.

The modular construction of the Crystal Palace, inspired by plants, enabled the forgoing of both pillars and load-bearing walls, thus rendering the entire inner surface available.

In the years following the Crystal Palace, the fascination of the *Victoria amazonica* continued to attract the interest of architects, and many ventured to construct buildings that were more or less inspired by its nervation. Among many others, there is Terminal 5 (part of which was once the TWA Flight Center) at John F. Kennedy International Airport in New York, designed by the Finnish American architect Eero Saarinen, and the extraordinary Palazzetto dello Sport in Rome, designed by the engineer Pier Luigi Nervi and the architect Annibale Vitellozzi in 1956. And it does not appear that the fascination for the plant with enormous leaves is going to stop: a few years ago, the Belgian architect Vincent Callebaut proposed the construction of floating cities, called Lilypads, that were completely autonomous and able to accommodate up to fifty thousand people, a design inspired in its shape by

In 1956, Pier Luigi Nervi and Anniballe Vitellozzi imitated the nervation structure of *Victoria amazonica* leaves in the construction of the Palazzetto dello Sport in Rome.

the great *Victoria amazonica*. Clearly the love story between this plant and architects is not over yet.

CACTI, WATER, AND SKYSCRAPERS

The prickly pear cactus (*Opuntia ficus-indica*) is a very common plant in many arid or semiarid areas in the world, where it manages to grow thanks to a series of adaptations to drought conditions. For a plant to survive in a desert environment requires uncommon skill: it has to be prepared to endure extreme heat that within the plant can reach temperatures above 70°C (158°F); it has to somehow get the water needed for survival from an environment where the average annual precipita-

The prickly pear (*Opuntia ficus-indica*) is a cactus native to Mexico but naturalized throughout the Mediterranean basin. Its structure is perfect for surviving with small amounts of freshwater.

tion is less than the amount of rain that falls in London on any day in April; and last but not least, it has to be able to defend itself against animals that want to eat it for dinner.

These may seem like impossible challenges, but not for the prickly pear and many other species of the Cactaceae family. Cacti do very well even in the hostile environments of the world's driest deserts, managing to bend prohibitive environmental conditions to their advantage by means of amazing metamorphoses that have transformed the very structure of the plant. In cacti, we see the most substantial of these mutations: the total disappearance of the leaves. Leaves are the venue for photosynthesis, but they are also the part of the plant through which much water is lost with the opening and closing of the stomata, tiny pores that allow the intake of carbon dioxide necessary for photosynthesis. By suppressing the leaves and transferring photosynthetic function to inside the stem, the prickly pear removes the main source of wasted liquids.

For all plants, the proper management of the opening and closing of the stomata is a problem that is not easily resolved. By keeping the stomata open, the plant's leaf receives the maximum CO_2 input and therefore achieves maximum photosynthesis. But these tiny and widespread openings (a tobacco leaf, for example, has about 12,000 stomata per square centimeter) facilitate the escape of water vapor. The solution for the cactus lies in knowing how to juggle between the different needs, implementing a policy of opening and closing according to the different environmental variables.

To make the most of droughtlike weather conditions, it is vital that a cactus's regulation of the stomatal opening be perfect: a minimal delay in closure during a particularly sunny day could lead to the collapse of even the most resistant plant. In other plant species, the acquisition of CO_2 and its transformation into sugars through photosynthesis takes place at the same time and always during the day. But in the Cactaceae this crucial process happens at different times of the day. And Cactaceae open their stomata at night, when more favorable environmental conditions ensure a smaller amount of water leakage, and turn it into sugar the next day, in the light. These remarkable plants have changed photosynthesis itself to meet the extreme demands of saving water.

Losing the least amount of moisture possible, however, is not enough; it is only one aspect of the problem. A certain amount of water must inevitably be consumed to ensure normal metabolic functioning. For the plant, then, it is necessary to find other sources to draw from that can compensate for the lost liquids. How can that be done in a land where it never rains? And, most important, how can Cactaceae suc-

The *Welwitschia mirabilis* is a gymnosperm (from the same group of plants as pines and firs) growing in the Kalahari and Namib Deserts, where they survive in conditions of extreme aridity.

ceed in an environment where the amount of water in the soil is zero? Many species of the *Opuntia* genus (the one the prickly pear belongs to) succeed in this seemingly impossible feat. Because of their amazing adaptability, these plants have learned to absorb water from the only source that can supply it in the desert: the atmosphere. The very thin hairlike spines that cover the cladodes, the structural elements of the prickly pear that are commonly known as pads, are not only a deterrent to animals but also an excellent tool for condensing atmospheric humidity. Moisture is trapped by the spines and conveyed in ever larger drops inside the cladodes, which, among their many functions, are also the main water reservoir for the plant. Similar systems to condense atmospheric water are used by numerous species, both plant and animal, through the unique structural features of their surfaces.

Namibia is an ideal place to see this. The Namib Desert is not only one of the most arid environments on the planet, it has also been like that for a very long time. Unlike deserts such as the Sahara, whose climate has undergone large swings between dry periods and wet periods in the last one hundred thousand years (there are even predictions of a return to a green environment—in "only" fifteen thousand years' time), the Namib has been relentlessly arid for at least eighty million years. This is a time span that is so long that it has enabled many species to evolve and adapt to its dryness by learning how to use the water from the fog that occasionally pushes inland from the ocean.

Among the species typical of this region are the *Welwitschia mirabilis* (or "platypus of plants," according to the famous moniker given it by Charles Darwin), which produces only two continuously growing leaves that can become sixteen feet long. This plant has adapted so well to extreme climates that it can live for literally thousands of years. Some specimens of *Welwitschia*, in fact, are well over two thousand years old, and its name in Afrikaans, *tweeblaarkanniedood*, means "two leaves that never die." The survival of this unique plant, described by the English botanist Joseph D. Hooker (1817–1911) as "the most exceptional plant

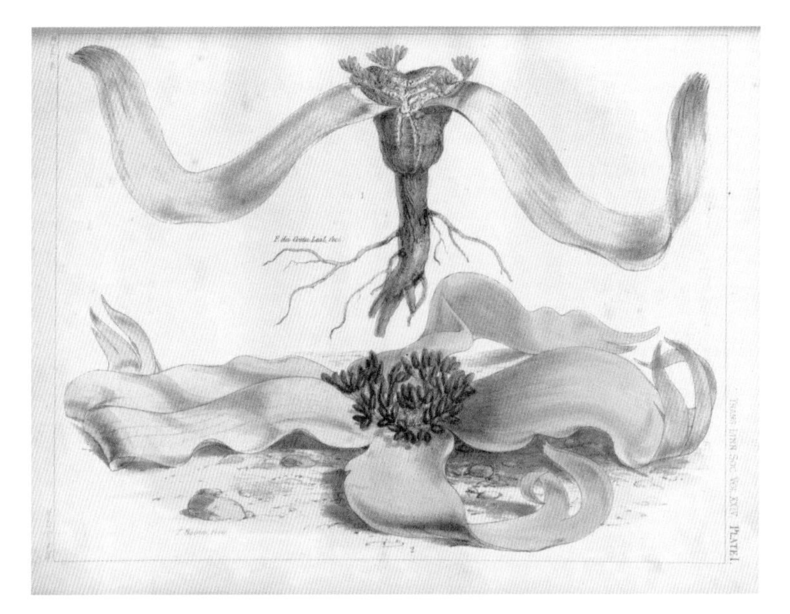

The *Welwitschia mirabilis* has a taproot that goes very deep and two leaves that can grow up to sixteen feet in length.

ever brought into this country, and the ugliest," does not depend on the length of its roots to gather water, as was long believed, but rather on the ability of its long, porous leaves to absorb water droplets produced by the atmospheric condensation of the ocean fogs resulting from the region's extreme temperature ranges.

The so-called fog beetles—insects of the Tenebrionidae family, endemic to the Namib Desert—have evolved similar mechanisms to collect atmospheric humidity. The *Stenocara gracilipes*, for example, positions itself at a forty-five-degree angle to the breeze coming from the sea and captures moisture from it thanks to wings that consist of alternating hydrophilic and hydrophobic surfaces: the water contained in the fog binds to the hydrophilic areas of the wing, forming droplets large enough to roll directly into the mouth of the insect. This mechanism has been imitated by engineers to produce fabrics that are able to absorb water from the atmosphere. Even thin structures such as cobwebs are able to collect moisture from the air.

Such techniques have been used throughout history by humans to obtain water in regions where it was scarce. The Italian architect Pietro Laureano has dedicated his entire professional life to the study of these traditions. He has identified the earliest evidence of such practices in the so-called solar tombs, special Bronze Age grave sites composed of a double circle and crossed by a corridor leading to a central hollowed-out space. Such constructions would have served as both places of worship and as conveyors of moisture. Similar stone structures, common in the arid areas of Puglia and Sicily, were used for the same purpose. Moist air enters the passageway between the stones, which has a lower tem-

Tiny drops of water from the condensation of moisture in the air, gathered by dandelion seed fibers.

perature inside because it is not exposed to the sun and is cooled by the hypogeum (an underground chamber) below. The resulting decrease in temperature causes the condensation of water droplets, which are then collected at the bottom of the cavity. During the night the process is reversed, producing similar results on the outer face of the stones. For many centuries these techniques, now long forgotten, provided the water supply for the populations of many areas of the Mediterranean and enabled the survival of humankind even in inhospitable regions such as the Sahara. Now, thanks to the work of people like Pietro Laureano, their use for purposes practical to our times is coming back into favor.

In this sense, the knowledge gained about the ability of Cactaceae such as the prickly pear to condense water has proved to be fundamental in designing ever more efficient and technologically advanced systems, which mimic the salient features used by plants. The skyscraper that will house Qatar's Ministry of Agriculture was inspired by the adaptations typical of the Cactaceae (remember that in that country the annual average rainfall is less than three inches). From the columnar shape common to many cacti to the opening and closing of the vents that guarantee air circulation inside the building, everything is designed based on lessons learned from studying the growth of plants in these arid areas.

Warka Tower, a prototype designed by the Italian architect Arturo Vittori, is another example of the technological and sustainable manifestation of such water collection and condensation systems. Even through the name we understand that this is a case of direct inspiration from plants: *warka* is the local name of a giant fig tree (*Ficus vasta*) endemic to Ethiopia—unfortunately increasingly rare—that constitutes a very important element for the local culture and ecosystem because it is appreciated both as a fruit tree and, due to its size, a gathering place for the community. Warka Tower has the stylized form of a tree (its splendid design was awarded the World Design Impact Prize in 2016),

An image of the Warka Tower designed by Arturo Vittori. It is a structure capable of producing water from atmospheric moisture condensation.

and, thanks to special nets specifically designed for high-efficiency condensation, it can produce up to twenty-six gallons of water a day from the atmosphere of an arid environment such as that in Ethiopia. In my opinion its low cost, high efficiency, ease of construction and use, and architectural beauty make it a perfect example of how brilliant innovators can revolutionize the forms and technology of our future by turning to plants for their inspiration.

Plants have always provided architects with solutions and inspiration. The majesty of the trees in a forest re-created by the columns of temples or the delicate grace of Corinthian capitals decorated with acanthus leaves are just some of the many examples that come to mind. It has been thousands of years since the Egyptians imitated the stem of the papyrus in the construction of the columns of the Temple of Luxor, and today plants are still a bottomless source of innovative ideas for the world of architecture. I hope that this trend will continue in the coming years: for one thing, it is difficult to produce eyesores if we are guided by the shapes of nature.

CHAPTER 7

LIVING WITHOUT FRESHWATER

Ocean: A body of water occupying about two-thirds of a world made for man—who has no gills.

—**AMBROSE BIERCE, *The Unabridged Devil's Dictionary***

Water is the substance from which all things originate; its fluidity also explains the mutations of those things themselves. This concept derives from the establishment that animals and plants feed on moisture, that foods are rich in juices, and that living things dry out after death.

—**THALES**

Praised be You, my Lord, through Sister Water, which is very useful and humble and precious and chaste.

—**SAINT FRANCIS OF ASSISI, *Laudes Creaturarum*, or *Canticle of the Sun***

PREVIOUS PAGE: **A glimpse of Nevada, a largely semiarid state that is fragmented by many mountain ranges.**

THE AVAILABILITY OF FRESHWATER IS NOT UNLIMITED

On May 21, 2005, at the graduation ceremony at Kenyon College, the American writer David Foster Wallace told the following story: "There are these two young fish swimming along and they happen to meet an older fish swimming the other way, who nods at them and says 'Morning, boys. How's the water?' And the two young fish swim on for a bit, and then eventually one of them looks over at the other and goes 'What the hell is water?' "

The problem of water today has a lot to do with that story. For those who live in most Western countries, water is so readily available, cheap, and, at least to our eyes, virtually inexhaustible that we cannot recognize its real importance. Even classic economic theory perceived the value of water more or less in the same terms. In 1817 the English economist David Ricardo (1772–1823) wrote in *On the Principles of Political Economy and Taxation*, "On the common principles of supply and demand, no rent could be paid for . . . the use of air and water, or for any other of the gifts of nature which exist in boundless quantity. . . . In the same manner, the brewer, the distiller, the dyer, make incessant use of the air and water for the production of their commodities; but as the supply is boundless, they bear no price."

In recent years, it has become increasingly obvious that due to growing demand, the scarcity of freshwater is becoming a threat to the sustainable development of human society. In its most recent annual report on global risks, the World Economic Forum mentioned the lack of freshwater as the most important threat in terms of potential impact. The first consequences of prolonged periods of drought are, unfortunately and dramatically, right in front of us. Research published by the University of California bears convincing evidence that the worst drought ever recorded since measuring instruments existed began in the winter of 2006 in Syria and the large area of the Fertile Crescent, where agriculture itself was born about twelve thousand years ago. It lasted for three consecutive years and was one of the main causes of the civil war in Syria. Such a long period of drought dealt a fatal blow to farming in an area already heavily limited in its productivity due to a chronic shortage of freshwater. It led to the migration of more than 1.5 million people from rural areas to the outskirts of large cities, with catastrophic consequences.

The available water on Earth is 97 percent seawater, which is useless for human consumption, as well as for agriculture and industry. The entire burden of water for human use rests, therefore, on the remaining

Soil dryness caused by persistent drought and climate change is becoming a planetary emergency.

3 percent. Considering that another 1 percent is unusable because it is in the form of ice at the poles, only 2 percent remains available, which must be used for the constantly increasing world population and the continuous improvement of their living standards, which requires ever more water to meet the growing needs of industrial production and irrigated agriculture. In absolute terms, globally and on an annual basis, there should be enough freshwater on the earth to meet this demand, but the fluctuations in time and space of water supply and demand are huge. Thus many regions of the world, at particular times of the year, suffer from water shortages. In a way, the essence of the problem is the geographical and temporal mismatch between the demand for freshwater and its availability.

The problem of supplying this resource will become increasingly important in the near future, as a result of the global population being expected to grow at least until 2050, when the planet will be home to about ten billion people, a remarkable number. These people will need freshwater both for their own consumption and, above all, for the production of food.

The extent of the problem becomes clearer when we realize that by 2050 we must be able to produce enough food to feed the additional equivalent to the entire population of the world as it was in 1960, that is, about three billion people. In other words, over the next thirty years we will find ourselves having to feed the equivalent of a whole new planet. Seen from this perspective, it is evident that such an increase in population could prove to be unsustainable for the planet if it is not accompanied by a drastic change in our production and consumption models.

Rendering the problem of needing to feed another three billion people even more serious, if that were possible, are some discouraging data on agricultural production. Over the past decade, despite a steady growth in yields worldwide, we have seen the worrying phenomenon of decreasing yields in more developed countries. One of the possible explanations, supported by numerous studies, is that agricultural yields, in many regions of developed farming, are approaching the maximum biophysical yield of the crop in question. This would seem to be the case for rice in China and Japan, grain in the United Kingdom, Germany, and the Netherlands, and corn in Italy and France.

Another factor is climate changes. A 2016 study by Navin Ramankutty, a professor of global food security and sustainability at the University of British Columbia in Canada, and his colleagues quantified clearly for the first time the overall cost of climate-related disasters in the second half of the twentieth century. By studying 2,800 hydrometeorological disasters, droughts, and extreme thermic events that occurred in 177 countries between 1964 and 2007, it emerged that such phenomena were responsible for reductions in cereal production amounting to around 10 percent (more than 70 percent of the calories consumed by humans derive from cereals).

Moreover, in developed countries, these phenomena have caused a reduction that is almost double that seen in less advanced agriculture. In Australia, North America, and Europe, harvest levels fell by an average of 19.9 percent due to drought, which is about twice the global average. The difference seems to be due to the greater uniformity of industrial crops in developed countries. In a way, this confirms the experimental evidence of the dangers related to the loss of crop diversification. Since all grains in North America are grown in huge areas in a totally uniform way in terms of species and methods of cultivation, if, for some reason, something unexpectedly damages a crop, the whole production suffers. In contrast, in most developing countries, grains come from a patchwork of different crops in small fields. If some of them are damaged, the others will usually manage to survive.

The increase in extreme weather events is one of the most obvious consequences of climate change, and all the projections suggest that in the near future we will face a further increase in their number and intensity. We must, therefore, expect further decreases in yields over the coming years.

Halophytes are salt-resistant plants that can grow using only seawater. Their study could be critical to understanding plants' resistance to salinity.

If the yields do not increase—and indeed in many cases, as a result of climate change, they will decrease—the only possible solution to meet the growing demand for food seems to be turning more land over to cultivation. This, however, presents many complications. Deforestation in favor of cultivation of food is no longer tolerable. It would destroy areas fundamental for the balance of our planet, just to get fields that would soon decrease their production potential and rapidly become completely sterile. The ends do not justify the means: the negative effect of deforestation on the climate—and consequently on crop yields—is much higher than the temporary increase in production due to the expansion of crops. Any policy aimed at resolving the food problem with deforestation and cultivation of huge territories would have catastrophic consequences for the whole planet.

In addition, much of the potentially arable land is not actually us-

The Cactaceae have evolved to survive in arid areas where the availability of freshwater is very limited.

able for various reasons, often compounded by humans. This is the case, for example, with soils that would be perfectly suited to cultivation if only they were not toxic with high concentrations of salt.

Soil salinity is a problem whose importance is little known. It is estimated that 8.9 billion of the 12.8 billion acres of arable lands used for agriculture around the world are salty. Nearly 10 percent of the Earth's surface (2,350 million acres) and 50 percent of all irrigated land (570 million acres) worldwide have problems of salinity. Global annual losses in agricultural production due to salinity are over $12 billion and constantly rising. Climate change causes rising sea levels, which in turn creates saltwater infiltration into freshwater aquifers and direct encroachment into coastal lands, ensuring a continuous increase in the extent of saline soils.

There are not as many easily cultivated lands as might seem at first glance. On the contrary, they are rare and much sought after, generating an increasing purchase of potentially arable land by governments concerned with ensuring their own food security. The growing phenomenon and its extent evoke deep apprehension. From 2000 to 2012 contracts for the exploitation of approximately two hundred million acres, representing more than 2 percent of arable land worldwide, were registered, mostly in African countries such as Sudan, Tanzania, Ethiopia, and the Democratic Republic of Congo. Large parts of Africa, Latin America, and Southeast Asia have suffered the same fate, and in recent years the phenomenon has also been extending to large areas of Europe.

Food security is the real problem of the twenty-first century. How to ensure enough food for a continuously growing population? How

Recently, due to man's handiwork and climate change, saline lands are increasing. Every year rising sea levels render large areas of fertile land salty, and therefore sterile.

can it be done considering that the availability of productive land and water resources is in dramatic decline? To respond to these urgent needs without drawing further on the planet's resources and without aggravating the already delicate climatic situation would require a revolution in our understanding of agricultural production. One possible solution that could meet our growing demand for food and that also respects the constraints imposed by environmental issues might be this: moving part of our productive capacity onto the oceans. This might seem, at first glance, pure science fiction, but if we analyze it carefully, we will see that there is nothing implausible about it. Ninety-seven percent of the planet's water is salt water, and two-thirds of the globe is covered in water; oceans, I have no doubt, will be our new frontier long before extraterrestrial lands will. To utilize them, we will of course have to overcome technical difficulties and increase the number of plant species in our diet, including plants that are more tolerant to salts. These are minor issues that we are more than capable of resolving.

LIVING OFF SALT WATER

An agriculture adapted to higher levels of salinity could represent a concrete response to the decline in freshwater availability in agriculture and favor cultivation in soils with high salinity. All conventional crops are sensitive to salt. Even the so-called tolerant species, which are in any case few, at best withstand an irrigation of freshwater mixed with 30 percent seawater. Concentrations above that result in sharp reductions in yield due to salt's toxicity to plants. For decades, a significant number of researchers have tried to improve salt tolerance of the more commonly grown plants, but unfortunately with very poor results.

One group of plants that have solved the problem of growing in areas of high salinity by themselves has become of interest in recent years. These are the so-called halophytes (from the Greek *alo*, salt, and *fito*, plant), native to naturally salty areas (salty deserts, coastal areas, brackish lagoons, and so on) that can grow and reproduce in soils that would kill any other species. The domestication and cultivation of these plants, many of which are edible by humans and animals, would allow the use of brackish water and seawater for irrigation and render coastal areas and any land affected by high salinity more productive. In addition, the study of the wide range of morphological, physiological, and biochemical adaptations that allow halophytes to resist salt might give us some possible solution to making common plants more tolerant.

Imagine if halophytes—or at least plants that are fairly tolerant of salt—could be grown on farms floating on the sea. Without concerns about space or water, the problem of food security could be solved forever.

JELLYFISH BARGE: THE FLOATING GREENHOUSE

A couple of years ago, I met Cristiana Favretto and Antonio Girardi, two young Italian architects with a strong interest in the plant world as a source of technical inspiration. This couple, united both professionally and personally, have dedicated themselves in recent years, and with very original results, to transposing some basic concepts from the plant world into the world of architecture. Shared interests, the visionary eccentricity of their projects, and the instinctive liking that the young couple kindled led us naturally to discuss our respective experiences and ongoing projects. That was how I found out about what Antonio and Cristiana called "Jellyfish." They were developing the idea of a floating greenhouse that could turn salt water into freshwater and use it to irrigate the plants inside. In their stimulating preliminary sketches, the greenhouse was made of a transparent dome from whose base branched off long ropes—needed to absorb the salt water—which, like tentacles, plunge into the water. Close up its shape so resembled the body of a jellyfish that calling it Jellyfish was almost obligatory for them. I found the idea not only fascinating but above all a definitive step forward in the direction of the sea farms that I had been thinking about for some time.

We began to examine the many technical problems to be solved and to wonder if it would be plausible to attempt to translate the idea into an operational prototype. I invited them to come see me at the LINV in Florence to take the discussion further and to find the most efficient way to collaborate. For weeks, thinking about how to turn Jellyfish into the greenhouse of the future, we added more and more requirements to what the tiny sea farm should be able to do. The end result was a fascinating and very ambitious project. Our intention was to create an autonomous system for the production of plant foods that would not require any farmland, would not consume even a drop of freshwater, and would be powered only by solar or other forms of clean energy, such as wind and wave power. We would not settle for anything less. We wanted to create a system that could produce food without consuming resources. Jellyfish would be our contribution to solving the problem of food security on our planet. So we added the word "barge" to jellyfish: Jellyfish Barge (JB) would be the lifeboat that would allow the production of food in the most catastrophic conditions.

Of course, producing without consuming resources is kind of the philosopher's stone of sustainable production. It is an almost impossible challenge. At the beginning, despite goodwill and the painstaking work of the team that we had formed, we could not meet all the requirements simultaneously. It was possible not to use freshwater, but only by using a lot of energy, and if we wanted to use little energy we could not do it with hydroponics (the growing of plants in a liquid nutrient solution). The more time that passed, the more it seemed that our visionary project was not feasible. Furthermore, to cover all the bases, we had imposed an additional constraint: each component of Jellyfish Barge had to be completely reusable and, where possible, obtained through recycling.

Growing lettuce on the Jellyfish Barge. The floating greenhouse enables the production of vegetables without the consumption of freshwater.

In the following months, the many issues on which we were working began to seem unsolvable. Every attempt we made came back with the same disappointing result. For a while we continued to go round and round in circles with the many aspects of the project. We were blocked by the numerous constraints and did not know how to resolve them—until we decided to go back to our original idea: get inspiration from the plant world; find in nature the solution to the technological problems that tormented us. Antonio and Cristiana redesigned Jellyfish Barge so that even structurally it followed the plant's basic model. The first step was to make it modular. Just as a plant consists of units that are reiterated according to its size, so Jellyfish Barge had to be able to work both alone (as a single autonomous floating greenhouse) and in ever larger groups suitable for the production of large numbers of plants. The basic form of the individual module became octagonal, the perfect geometric figure for good management of space and that ensured the preservation of open spaces for transport or routine operations when several modules were joined together.

Nature and plants also inspired the design of the component that gave us the most trouble: the seawater desalination system. The words with which, in the *Codex Atlanticus*, Leonardo gives a succinct description of the water cycle came back to mind: "So therefore, one may conclude that the water passes from the rivers to the sea, and from the sea to the rivers." If we think of the natural cycle of water, we realize that it is a powerful desalinator. When water evaporates from the sea, it leaves its salts behind. Therefore, when water vapor, formed into clouds, condenses and falls to earth in the form of rain, it is freshwater. Through evaporation powered by the sun, massive amounts of water are desalinated every day. Plants participate in this natural cycle through the transpiration of water by their foliage. Forests such as the Amazon rain forest transpire in such quantities as to significantly affect the formation of the earth's climate, and some trees, such as mangroves, are able to transpire seawater directly.

Thus in solar desalination we had found the most suitable system to produce the freshwater required. It is an astoundingly simple process: water evaporates under the action of the sun and then transitions into a liquid state through condensation in a cooler environment. The process, as we discovered in the course of our work, had been used by US soldiers during World War II to produce freshwater even in the most desperate situations. The US Army had even produced a special kit that, exploiting the rays of the tropical sun, was able to derive from seawater all the drinking water necessary for personal survival. Before long, we were able to design desalinators that, using solar heat, were able to produce at Mediterranean latitudes more than fifty gallons of freshwater per day, more than enough for the needs of the hydroponic system that would enable the growth of plants in the greenhouse.

Having resolved the problem of water, we were ready for the construction of the first prototype. But to move forward, we had to find a patron who believed in the project. That proved to be easier than expected. Everybody liked Jellyfish Barge, and the importance of producing food without using valuable resources was more than evident. A prominent foundation in Florence became our main financial supporter.

In a short time we created the first working prototype. Everything was going perfectly: the greenhouse floated, the hydroponics system held, water was derived from the desalinators in the quantities needed. The only downside was the quality of the water, which was too pure for our purposes. The water produced due to solar desalination is, in fact, comparable to distilled water and does not contain any mineral element whatsoever. To avoid this complication and at the same time increase

The first prototype of Jellyfish Barge was entirely made of wood and was able to produce vegetables without using freshwater, soil, or energy other than that of the sun.

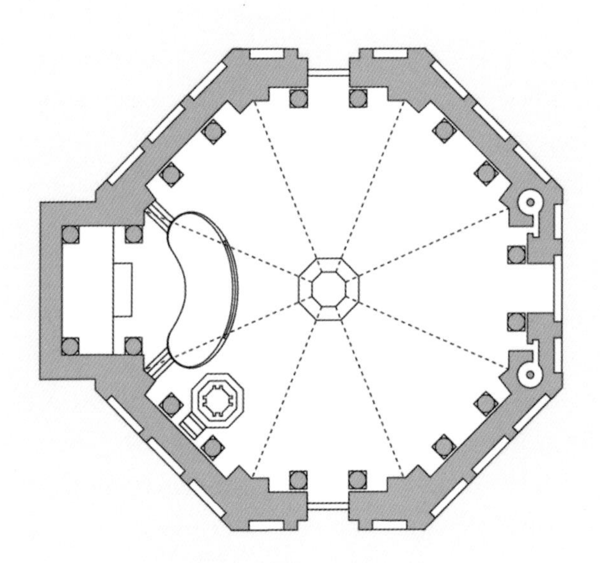

The octagonal floor plan of a Romanesque building. Castel del Monte in Puglia, the baptistery of Florence, and Thomas Jefferson's home Monticello in Virginia are among the best-known examples of the octagon design.

the usable stock, we mixed the water produced by the desalinators with 10 percent of that taken from the sea. In this way, it was enriched with the minerals contained in seawater, without creating toxicity for the crops. Jellyfish Barge began to function brilliantly and grow vegetables at full speed. In a month the floating greenhouse produced about five hundred heads of lettuce ready for consumption, and many observers realized that growing vegetables without consuming any resources was no longer a pipe dream. A visionary idea had found its practical application. Plants grown in the greenhouse were our contribution to a sustainable future.

Jellyfish Barge was one of the Italian projects presented at the Milan Expo in 2015, and hundreds of thousands of people were able to see it float and visit it in the former dock of the Lombard capital. It has been exhibited in many cities around the world and won a great many international awards, some of which are very important and are promoted by the United Nations. It is also really beautiful, as evidenced by the architectural awards it has won; yet it does not seem to interest investors. Even if it succeeds in the apparent miracle of producing vegetables without consuming any resources, the market does not seem to care very much. As a colleague of mine once said, not without a bit of satisfaction, "Either projects win prizes or they go on the market." And it looks as though JB, at least for the moment, is destined only to win awards.

But we are not discouraged; sooner or later, inevitably, we will have to cultivate the sea to produce food. And Jellyfish Barge is already up and running, ready for the task.

The hydroponic system inside Jellyfish Barge increases the greenhouse's efficiency in producing desalinated water.

CHAPTER 8

FROM PLANTS TO PLANTOIDS

> Look deep into nature, you will understand everything better.
>
> —**ALBERT EINSTEIN**

PREVIOUS PAGE: A single root system may be composed of billions of apexes. The image shows the intricacy of a small portion of the corn root system.

IS THE BIOINSPIRED APPROACH NEW?

After years of premature announcements, concerns, corrections, and clarifications, the long-awaited robotic revolution seems to be meeting its challenge. Cheap and reliable robots are replacing men and women in many tasks that until a few decades ago could be done only by humans. Some are already part of our daily life: robots that can vacuum apartments, mow lawns, or clean up litter from streets are no longer the preserve of science fiction movies.

Despite the undeniable reality that robots have now become indispensable in many fields, the general perception is that their proliferation, as feared by some as it is hoped for by others, is still a long way off. This is a largely mistaken perception, tied to the ideas we have of such machines. In truth, their diffusion has increased exponentially: in fields such as industrial automation, medicine, underwater research, and the like, their use has become widespread and irrevocable. Every day we hear about new applications: bomb squad robots, car driver robots, medical diagnostic robots . . . Yet, based on our daily experience of the world, few people seem to feel that there are more automatons now than there were, say, thirty years ago. Why is that? My personal opinion is that this perception is linked to the popular idea of robots, formed through the hundreds of movies and novels on the subject; an idea that corresponds to that of an android, constructed in such a way as to mimic the features and characteristics of a human.

The word *robot* comes from the Czech word *robota*, meaning a difficult or unnatural work (in Polish the word *robotnik* means "worker"). It was first used by the Czech writer Karel Čapek in a 1920 play called

R.U.R. (*Rossum's Universal Robots*), which became an immediate hit, thus encouraging the spread of its underlying theme. However, in Čapek's science fiction drama, the artificial workers who are supposed to simplify the life of man are in fact clones, that is, organic humanoids. Maybe that is why the belief that a robot, even if mechanical, is essentially a humanoid slave, a simplified replication of us, became so widespread so quickly. A few years afterward, Fritz Lang's 1927 masterpiece, the Expressionist film *Metropolis*, fixed forever in the collective imagination the image of a man-machine. But, outside of fiction, who says that the human form is the most suitable for a robot?

Our preconception that these machines must necessarily resemble humans is revealing, because it suggests that our approach toward the production of new technology is that of a replacement, expansion, or improvement of human functions. Humans have always attempted to replicate themselves—or at least their basic animal design—in the construction of their instruments. Consider that the very symbol of modernity, the computer, is based on ancient schemata: a processor, representing the brain and that has the function of governing the hardware, plus video and sound cards, which translate our organs into something technological. Everything that humans design tends to have, in a more or less overt way, the same common basic architecture, with a "thinking brain" that governs the "actuating organs." Even our societies, as we have seen, are built on this model. Fortunately, in recent years, the so-called bioinspired approach—the one that looks to nature as a role model for solving technological problems—has started to be used to design and produce new materials and machinery.

Bioinspiration has brought with it a breath of fresh air even in contemporary robotics. Man is no longer the only model to inspire: the entire animal world has become a source of solutions to study and imitate. In recent years animaloids and insectoids have gained a stronger and stronger foothold, and robots replicating salamanders, mules, and even octopuses have been successfully designed. After all, if you want to build

TOP: Scenery from the science fiction drama *R.U.R. (Rossum's Universal Robots)* staged in Prague in 1921.

BOTTOM: The office of Domin, the play's protagonist demiurge, in Vlastislav Hofman's drawing.

a robot that grabs and moves objects under water, taking inspiration from the lively tentacular intelligence of the octopus is certainly a good idea. And if you have to design an amphibious robot that can easily move from an aquatic environment to a terrestrial one, what could be better than a salamander? However, for now bioinspirations seem to be confined to the animal kingdom. And plants? Well, so far no one has thought that they could make a significant contribution to the issue.

I believe there are many good reasons for our emerging technology to mimic the plant world. Plants consume very little energy, they carry out passive movements, they are constructed in modules, they are robust, they have a distributed intelligence (as opposed to the centralized one of animals), they behave like colonies. When you want to design something robust, energetically sustainable, and adaptable to an environment of continuous change, there is nothing better on earth to use as inspiration.

WHY PLANTS?

Perhaps you are wondering how robots inspired by plants could actually be useful. Well, let me recapitulate. Plants are multicellular photosynthetic organisms characterized, albeit with some exceptions, by an elevated portion and a root system. To offset their sessile (attached directly by their base) nature, and to adapt to changing environmental conditions without being able to move, they have developed the ability to move through growth, showing extraordinary plasticity.

The set of responses to the environment that manifest themselves in a movement is commonly known as tropisms. These are characterized by a marked directional growth of the organs, especially of the roots, in response to external stimuli, the main ones being light (phototropism), gravity (gravitropism), contact (thigmotropism), humidity gradient (hydrotropism), oxygen (oxytropism), and electric field (electrotropism). To these, recently and based on the studies carried out at my laboratory, was added the so-called phonotropism, that is, growth regulated by a sound source. The combination of these mechanisms allows the plant to survive in hostile environments and colonize the soil through the creation of a root system that ensures its survival and stability (and which is often far greater than the foliage in mass and length, being able to reach dimensions that are hard to imagine).

To dramatically increase the absorbent surface of the roots, nature has resorted to a trick similar to the one attributed in classical poetic narrative to Dido, the legendary founder of Carthage. It is said that the African king Iarbas had granted the queen and her people, exiled in Tyre, all the land that was possible to cover with a cowhide. It was, obviously, a trick. However, the future queen of Carthage knew how to resolve the situation in her favor: she cut the leather into very thin strips and, binding them together, succeeded in surrounding the hill on which she wanted the city to be built. Similarly, a single plant of grain can achieve a linear development of more than twelve miles, if you take into consideration the total length of root hairs. If, on the other hand, we measure the volume of those hairs, we realize that they can fit into a cube that is little more than a half inch in width.

Another fundamental characteristic of root apexes is their ability to find a way to grow even in very resistant surrounding matter. Despite their fragile appearance and delicate structure, they are able to exert extraordinary pressure and can break even the hardest rock thanks to cell division and expansion. Because roots can grow, in fact, the size of the pores or cracks in the ground must be greater than the size of the root apex. In this way, the water within the apex's cells can generate a turgor (rigid, swollen state) that gives it the strength required for elongation and growth. The osmotic (fluid diffusion) potential of a root creates a gradient that sustains the absorption of water into the cells and these, swelling, push their cell membrane against the rigid wall. Depending on the species, pressure exerted in this way can be significant enough to enable roots to break highly resistant materials such as asphalt, concrete, and even granite.

INDIVIDUALITY IN PLANTS

Another little-known feature of plants from which robotics could draw inspiration, is their reiterated modular construction. The body of a tree consists of repeated units that together constitute its general architecture and define its physiology. This is quite different from what happens in the animal kingdom. Absurd as it may seem, even the definition of "individual" that we use for animals has little relevance to the world of plants. There are at least two different definitions of "individual":

1. *Etymological:* An individual is a biological entity that cannot be divided into two parts without at least one of the two dying.
2. *Genetic:* An individual is a biological entity that has a stable genome in space and time. In space because it is the same in any cell in the organism; in time because it extends across its whole life span.

It is easy to see how such definitions have little sense when applied to almost all plants. Let us start with the etymological question: a plant, if split into two, multiplies. At the end of the nineteenth century, the French naturalist Jean-Henri Fabre (1823–1915) wrote that "for animals, in the vast majority of cases, division means destruction; for plants, to divide is to multiply." This is a notion that is clear not only to scholars of plants but also to enthusiasts: nursery production, based on propagation through cuttings or by grafting, uses precisely this quality.

But even genetic stability seems to be less important for the plant world. In an animal, whatever its size, the genome is stable in every cell and for the whole of its life. In plants this rule does not seem to be of value, as anyone who has studied the so-called gemmarian mutation (changes to the DNA of the buds) in fruit trees knows well. In the history of fruit arboriculture, in fact, "mutant" branches have often been identified on a tree whose fruits were of special interest. There are plenty of examples, given that many varieties originated in this way: nectarines almost certainly originated from a gemmarian mutation of the peach; pinot grigio is a gemmarian mutation of pinot noir. Another fascinating example of different genomes that coexist in the same plant is the so-called chimeras, that is, individuals that—like the monsters of Greek mythology—consist of different specific traits, stemming from the parts of a graft that developed together. The numerous "quirks" common to many fruit species, such as the orange or the vine, are a prime example of this peculiarity of plant life. And among the many chimeras, the famous Bizzaria of Florence at least deserves a mention. This is a very rare variety of citrus with the peculiarity of producing fruit that has aspects of both the bitter orange and the citron, distributed irregularly. This plant, for a long time pride of the Medicean collections and described for the first time in 1674 by the then director of the Pisa Botanical Garden, Pietro Nati (1624–1715), was long considered extinct and was rediscovered only in the 1970s. Curious examples aside, though, it is easy to find similar genetic differences in all trees of a certain age.

It seems difficult to define a plant as "an individual"—so much so that at the end of the eighteenth century the idea that plants—and in particular trees—could be considered real colonies, consisting of reiterated architectural units, began to circulate. In 1790 Johann Wolfgang von Goethe wrote, "Lateral branches that originate from the nodes of a plant can be considered as separate young plants that attach themselves to the body of their mother, in the same manner in which the latter attaches itself to the ground." And Erasmus Darwin, taking up Goethe's thinking, stated in 1800, "Every bud of a tree is

One of the many citrus chimeras, the famous Bizzaria. Today you can admire it in the Medicean Villa di Castello and Giardino di Boboli collections in Florence.

an individual vegetable being; and that a tree therefore is a family or swarm of individual plants." In 1839 his nephew Charles added, "We may consider the polypi in a zoophyte, or the buds in a tree, as cases where the division of the individual has not been completely effected." Finally, in 1855, the German botanist Alexander Braun (1805–1877) observed, "The view of plants, and especially that of trees, suggests that this is not a unique and individual being, like an animal or a man, but rather a collection of individuals assembled together." The concept of a "plant colony" has long had distinguished advocates and also implies the idea—very interesting for any robotic application—of increased longevity: the colony survives its components; the single polyp lives for a few months only, while the coral that hosts it is potentially immortal. The tree demonstrates something similar: the basic architectural unit has a short life, while the colony (the tree) could live virtually forever.

To this we can add that the concept of a reiterated unit applies not only to the external part of the plant but also to the root system. Every single root has, in fact, its own self-contained command center that guides its direction but that, as in a real colony, cooperates with the other root apexes to resolve problems related to the life of the plant. Plant evolution has led to the development of a distributed intelligence—that is, a simple, practical system that enables them to find effective responses to the challenges of the environment in which they live. It is a remarkable achievement.

A single tree can be considered a colony of architectural modules that repeat themselves.

THE PLANTOID: AN EXAMPLE OF PLANT BIOINSPIRATION

As we have seen, there are numerous factors that support drawing inspiration from plants in the construction of robots. Convinced of this, I started to develop the idea of "plantoids" in 2003. The word *plantoid* seemed appropriate for this new type of automaton because of its similarity to *android*. I dreamed about machines that could be useful in countless ways, from the exploration of soil to the exploration of space. Since my knowledge of robotics was, and continues to be, very limited, it was obvious that alone I would never be able to turn my vision into something tangible. In fact, I was afraid, as is often the case in academia, that this idea would remain unrealized, locked in a drawer. Luckily, that is not what happened. During the period when I was talking endlessly about building plantoids with anyone who was so unwise as to stop and converse with me (I have noticed that if something interests me, I tend to be monomaniacal, a quality my friends and colleagues know all too well), I encountered the perfect person with whom to transform into reality what had hitherto been only a topic of theoretical discussion. When I met Barbara Mazzolai, who today directs the Center for Micro-BioRobotics at the Italian Institute of Technology in Pisa, she was already an outstanding robotics researcher and, thanks to her university education, was also extremely knowledgeable about biology. Talking to her about automatons and plants became a habit. The idea of plantoids excited us, and the more we discussed the possibilities for building them, the more we were convinced that it was achievable. There were many technical problems to deal with, of course, but we were confident that they could be solved. The plantoid had to see the light of day!

To build a robot from scratch—especially if it is a totally new concept—that really works and is not merely a mechanical toy is

An artistic representation of plantoids. Inspired by the workings of plants, these robots can be used in all cases requiring soil exploration, from the search for resources to that of contaminants.

something that requires time, effort, and money. Like all enthusiastic researchers, we were ready to commit our time and our effort, but money was simply not something we had (I am too embarrassed to tell you what an Italian researcher's salary is). To identify an institution, a foundation, or another organization that would be willing to embark on this venture, paying all the expenses, was not easy. It took a long, long time. At first, the validity of the theoretical basis and the soundness of the system, which to Barbara and me seemed crystal clear and without any weak points, left our interlocutors completely cold (or at best lukewarm). Unfortunately, as I have seen many times, it is not easy to convince those who have always seen plants simply as borderline inorganic organisms that are good for decorating our gardens of their truly extraordinary abilities. It was even more difficult to persuade our lenders that, by imitating plants, we could build a new generation of robots.

But you must never despair when you go hunting for funds to build a visionary project: if you really believe in what you are proposing, eventually someone will share your enthusiasm. And that is what happened with Ariadna, an initiative of the European Space Agency's Advanced Concepts Team. Our arguments about the possibility of using inspiration from plants to build robots to be used in space exploration convinced the people there, and without delay they financed what is called a feasibility study. The investment was limited and did not allow us to build anything, but it really helped us firm up our ideas and anticipate the problems that could arise in the construction of a plantoid. In the end, we compiled a sizable document for ESA titled *Bio-Inspiration from Plants' Roots*. It went into great detail on the construction plans for a plantoid and on its possible uses for the exploration of space (and of Mars specifically). Our underlying thesis was simple: because plants are the pioneer organisms par excellence, by studying their systems of survival and replicating them in a plantoid, we could build a machine with a greater ability to survive in hostile environments. And nothing is more hostile than an extraterrestrial environment such as that of Mars. The project envisaged the transport of a large number of plantoids into Mars' atmosphere, where they would be released. Little more than four inches high, these plantoids would be scattered across the red planet, instantly opening up and pushing their roots into the soil. They would explore the subsoil with their roots, while a number of simulation leaves on the surface would feed energy to the robot forever thanks to photovoltaic cells (those with the ability to convert light into electricity). Our project represented a complete overturning of how exploration of Mars was being envisioned. Instead of continuing to send huge and very expensive robots, which move very slowly and explore only tiny areas, we would send thousands of plantoids. These, exploding in the atmosphere like seeds, would be scattered across a vast area and, without moving but

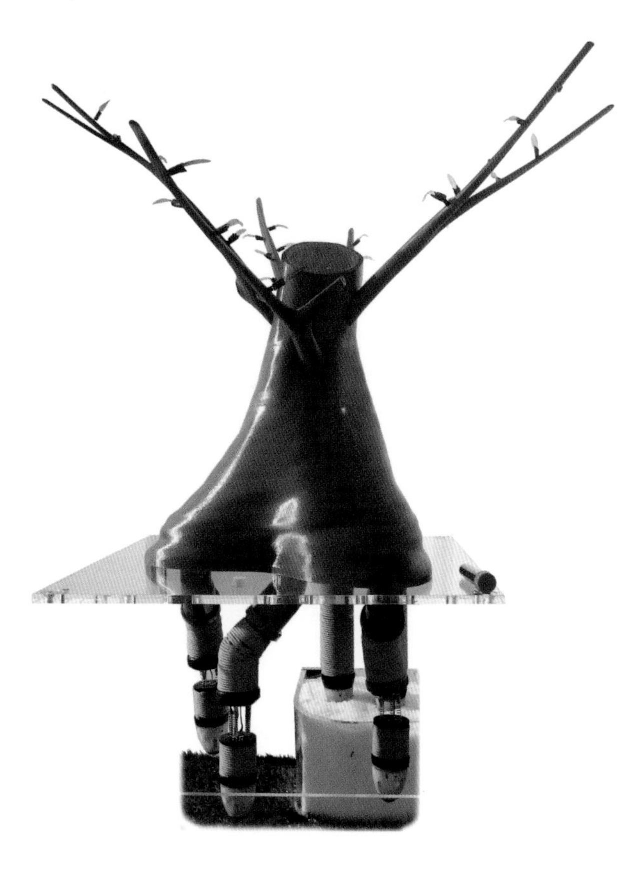

The first plantoid prototype built within the European FET program, able to grow its root apexes in the soil.

communicating with one another and with earth, they would transmit data on soil composition—data so comprehensive and so accurate as to give us a range of continental mapping.

Once we had finished the study for ESA, the project came to a standstill again, and for years no one was willing to fund it—until, in 2011, Barbara and I tried to get a contribution from the European Union through an initiative that awarded more "visionary" ideas, ones with a high risk of failure but also with a high level of innovation. The Future and Emerging Technologies (FET) program was, and still is, the arena in which the most revolutionary European technological projects compete for adequate funding. Our proposal, entitled "Plantoid: Innovative Robotic Artifacts Inspired by Plant Roots for Soil Monitoring," achieved an astounding score of 15/15—the maximum possible. And so it was awarded suitable financing, which would allow us finally to build the first plantoids.

The next three years were devoted to resolving the thousands of problems posed by the design and construction of the many modules that make up a plantoid and, ultimately, by its final realization. One of the biggest stumbling blocks for Barbara's laboratory was how to imitate the growth of roots. It was not a minor issue: creating autogrowth mechanisms remains one of the most difficult objectives to achieve in robotics.

In plants, root growth and movement processes are implemented essentially by two means: cellular division in the apical meristem, immediately below the tip of the root, and cellular elongation of an area behind the apex, called the elongation area. Both mechanisms were imitated in the construction of our robotic root apexes through the use of a plastic tank that promoted the growth of the robotic root. Furthermore, with the intent of replicating the different sensory capacities of the root, our robotic root apex was given an accelerometer, to replicate the ability to follow the direction of gravity; a humidity sensor, to mimic the ability to perceive minimal gradients of water; a number of chemical sensors; osmotic actuators (special devices capable of transforming osmotic pressure into movement), which ensured the direction of and penetration into the soil; a microcontroller, which handled the information resulting from the different sensors and re-created the distributed intelligence of root systems. Having built the robotic roots of our plantoid, all that was left to deal with were the leaves. That problem, which was less complex, could be resolved with photovoltaic cells that would imitate the photosynthetic process and provide the energy necessary to carry out all operational functions.

Replicating the adaptive strategy of the plant, the plantoid moves very slowly in order to efficiently explore the environment and to allow elevated powers of implementation and low energy consumption. The apex of the plantoid grows and moves in the ground at the same time as it communicates with all the others, collecting data and making it possible to use distributed intelligence strategies typical of the plant world.

Today plantoids are a reality. They can be used in the most diverse of circumstances, from instances of radioactive or chemical pollution, terrorist attacks and the mapping of minefields to space exploration, prospecting for oil, and inventing Agriculture 2.0.

Barbara continues to improve them and to modify them for particular applications. We are only at the start of this very interesting journey, but already the number of people who believe in the technological possibilities offered by plant bioinspiration is increasing. I hope—indeed, I am convinced—that it will not be long before we see groups of peaceful plantoids taking care of our gardens and farms.

CHAPTER 9

SPACE PLANTS

A blade of grass is a commonplace on Earth; it would be a miracle on Mars. Our descendants on Mars will know the value of a patch of green.

—**CARL SAGAN**, ***Pale Blue Dot: A Vision of the Human Future in Space***

VEGGIE
Vegetable Production System

PREVIOUS PAGE: Veggie is a minigreenhouse designed by NASA for growing plants in zero gravity.

OUR TRAVEL COMPANIONS IN SPACE

"The man (or woman) who will first set foot on Mars has already been born." For some years now, in space agencies around the world, this has become a kind of mantra repeated at every turn. There is no discussion, interview, or conference concerning the future of space exploration in which someone does not feel the need to remind us that the Martian Neil Armstrong is already among us. Whether it is true or not, I cannot say. Another cliché that haunts anyone interested in space research is that there are no technical difficulties in putting a man on Mars that cannot be resolved: we have been ready for this memorable feat for some time. Yet forty years have passed since we sent a man just to the moon. The recently deceased American astronaut Eugene Cernan, who, on December 14, 1972, climbed back into *Challenger*, having landed on the moon three days earlier, to travel the 240,000 miles back home, was the last man to visit our beloved satellite. Over time, he has become as famous as Armstrong, who was the first to walk on the moon on July 21, 1969.

The moon is right here in front of us, ridiculously close compared to the thirty-four million miles that separate us from Mars at the closest point of our respective orbits (a point that is arrived at every twenty-six months, when the two planets are in "conjunction"). It is likely that it is not so much the technical difficulties that are slowing our conquest of the solar system as the economic ones and conflicts about the priorities of research in this field. However, we can be sure of one thing: whatever destination—near or far—we choose as the next step of our expansion into space, we cannot go there without plants. Yet we tend to forget this. Indeed, you could say that we tend to dismiss an indisputable assumption: we humans are totally dependent on plants. The food and oxygen we consume are products of the plant world. Without plants, life would not be possible. If we think about it with a cool head, it is clear that this is a real dependence, which severely limits our ability to move in the universe. We need to have a clear awareness of the fact that plants are the engine of life. But instead we are suffering from a persistent and unexplained plant blindness. If we take it as a given that a scuba diver cannot operate underwater without oxygen in his cylinder, why can we not understand that in the same way our species depends entirely on the plant world? If we want to move somewhere away from Earth—even just a few thousand miles into orbit—we will need a good supply of plants.

Anyone who has read Andy Weir's novel *The Martian* or seen Ridley Scott's 2015 film adaptation with Matt Damon as a botanist-astronaut will immediately understand what I mean. The ingenious Mark Watney, given up for dead by his colleagues, stays alive by cultivating potatoes using the soil on Mars. The importance of plants for food and oxygen is obvious, but other reasons for the fundamental role of plants in any long-term space mission are less so. One of these is the positive effect that plants have on humans' psychic balance.

The human factor is one of the most significant among the many issues to be resolved before embarking on long journeys into space. A journey to Mars would require an estimated time that varies from six or seven months to about a year (depending on a whole host of elements, including how much fuel we can take). It would take just as long to return to Earth, and a few months (probably more than a year) would need to be spent on the red planet waiting for the orbits of Earth and Mars to be once more in conjunction. The math is easily done: it is a journey with a total duration of between two and three years. Just imagine having to stay closed inside a box of a few square meters, bristling with sharp edges and tools, with very little space, in the company of three or four other crew members, with no possibility of privacy in the middle of the most infinite space and in zero gravity for that amount of time.

In the simulations carried out on Earth, under conditions that attempt to replicate those of the Martian journey, crews have manifested a problematic tendency to mental deterioration after a few months of such treatment, even though they were chosen from thousands particularly for their strong nerves. The human factor is the real hurdle to overcome in selecting crew members who possess both the required technical skills and the capacity to handle the psychological and mental pressures of deep-space travel. Teams of experts have been working on the issue for years, and one of the solutions that seems to offer the best support is to equip the mission with a good supply of plants.

The beneficial effects that the presence of plants has on the human

The University of Florence laboratory, where, my collaborators and I do much of the research presented in this book.

mind have been known for decades. People with mental disorders find relief through their relationship with plants in the countless horticulture therapy centers spread all over the world. School-age children with attention deficit disorder have shown significantly better performance in their studies in the presence of plants. A decade ago, the LINV, the laboratory I run, published research on the subject. At the time, we put a number of second- and fourth-grade school children (children of seven and nine years old) through a series of attention tests, to be carried out in places both with plants and without them: in a classroom with windows that did not look out over greenery and in the school garden with trees. Even though the standard classroom guaranteed an environment that no doubt was better suited to concentration (no distractions, no noise), the results obtained in the garden classroom, in the presence of plants, were far better.

In 2014, aboard the International Space Station, plant cultivation was initiated in a minigreenhouse called Veggie, which enabled the growing of not only lettuce but also, in January 2016, the first flowers grown in the absence of gravity (zinnias, to be precise). Raymond Wheeler, the director of the Advanced Life Support activities at NASA, acknowledged that such experiments have produced noticeably positive effects on astronauts' mood. As a result, the search for bioregenerative life support systems (BLSSs) in space has intensified. These artificial ecosystems mimic the interactions among microorganisms, animals, and plants typical of terrestrial ecology, in which the waste products of each one become resources for another. In these modules, the role of plants is critical as they produce oxygen, remove carbon dioxide through the process of photosynthesis, purify water through transpiration, and, finally, of course, yield fresh food. It is fascinating to think how human space exploration, which for us has always been one of the pillars around which we imagine the future, is inextricably linked to an activity as old as agriculture.

For decades, it was impossible to find a botanist among the staff of a national space agency, but in the past twenty years things have changed. Even the most intransigent technocrats have had to admit that the presence of plants provides a serious advantage in our efforts to explore and colonize space. Of course, the space environment differs from that of Earth because of the different gravitational conditions and the greater impact of cosmic rays. But plants grown in space in conditions of weightlessness, even if there may be problems of chromosomal aberrations or changes in the biological cycle, can normally adapt themselves. In general, microgravity, as well as conditions where the force of gravity is greater than that on Earth (hypergravity), are serious sources of stress for plants. However, unlike other stressful factors such as drought, extreme heat, salinity, anoxia (lack of oxygen), and many others with which the plant world has dealt in the course of its evolution, the absence of gravity is unprecedented for every organism born on Earth. Gravity is a fundamental force that influences every biological (but also physical and chemical) phenomenon present on Earth. The physiology of organisms, their metabolism, their structure, the way they communicate, the shape itself of any living being, are all molded by the extent of this force.

To study the effects of variations in gravity on plants, the European Space Agency conducted research with the International Space Station, the Bremen Drop Tower in Germany, and the supercentrifuge at Noordwijk, the Netherlands, as well as parabolic flight and sounding rockets. The drop tower is a 480-foot tower built by the University of Bremen, in which it is possible to carry out experiments in free fall (under conditions similar to the absence of gravity) for a duration of five seconds. The sounding rockets are real missiles launched from a base in Kiruna, Sweden, that can host zero-gravity experiments lasting up to forty-five minutes inside them. The ESA's supercentrifuge at Noordwijk is a huge centrifuge that can accommodate experiments weighing hundreds of pounds. With this equipment, it is possible to simulate the effect of forces of gravity that are higher than those on

The supercentrifuge at the ESA center in Noordwijk, the Netherlands, enables hypergravity experiments that have a high total weight.

Earth, to which plants could be subjected during moments of acceleration in any space travel.

Over the years my laboratory has used each of these means to study the effects of variations in gravity on plant physiology. An LINV experiment, intended to identify the principal genes involved in signaling stress activated in the absence of gravity, had the honor of being included on the journey of the Space Shuttle *Endeavour* on May 16, 2011. The results obtained allowed us to formulate the hypothesis that changes in gravitational acceleration constitute a stress on plant physiology. The good news is that, similar to what happens under more traditional stress, it is possible to acclimatize plants to increase their tolerance of the variations in gravity.

The Drop Tower at the University of Bremen is an extraordinary scientific facility that enables the study of the effects of short periods of microgravity.

FLYING FREE

I have always loved space research and the fascinating world of engineers, scientists, visionaries, and madmen who revolve around it. So when, in 2004, the ESA accepted our proposal for a series of experiments requiring a parabolic flight campaign, my first thought was: I am about to access one of the world's most exclusive clubs, one with a very select number of members who have experienced the absence of gravity and, certainly, the one that more than any other I had fantasized about being a part of as a boy, when I devoured science fiction novels at a blistering pace.

To take part in a parabolic flight, you have to arm yourself with considerable patience and undergo a long medical and bureaucratic process: forms, requests, tests, approvals, medical examinations, more tests. But it is worth it. I remember very clearly every moment of my first campaign—I would go on to participate in another six—aboard the Airbus A300-ZeroG, the modified plane that ESA uses for parabolic flights, departing from Bordeaux-Mérignac in France.

The week before the flight, our Italo-German team had organized on board the tools and equipment we had decided we needed for the experiment. We wanted to study the first cellular signals from the roots of corn seedlings at the very moment they found themselves in zero gravity. The experiment was very complex. It required the measurement of faint electrical signals that we thought would be produced in a specific area of the root apex (which is, we must remember, a sophisticated sensory organ) that measured less than one millimeter in diameter, in the first few instants of exposure to zero gravity. The unknowns were innumerable. We had no idea how the aircraft's vibrations would affect the delicate measurements. We did not know if the plants would remain in a good enough state of health during the flight to be able to respond to the different gravities with the promptness required. We did not know the conditions we would face, nor did we know if we would be able to replace the plants during the experiments. In essence, we were totally unprepared to work in those unprecedented experimental conditions. My personal opinion, though I would never have admitted it, was that the experiments would be a failure.

Among the many reasons for concern, one particularly troubled all the novices we got to know in the preceding week. Parabolic flights are notorious for the deleterious effects they have on the stomachs of the participants, resulting in the flights being affectionately nicknamed "the vomit comet." But that did not worry me too much. I have never suffered from seasickness—I thought naively—and certainly a banal stomach problem was not going to stop me from conducting the experiments and enjoying my first astronautic experience. After a mostly sleepless night imagining everything that could go wrong, the fateful day arrived and with it the coveted blue ESA jumpsuit supplied to participants of the flight. I felt like a real astronaut when I put it on. I was totally indifferent to the fact that it was a couple of sizes too big for me. It had everything I needed: it was blue, with a flashy badge of the space agency and the inscription "Parabolic Flight Campaign." The generous number of airsickness bags that filled most of the ten pockets of the suit simply provoked a slight smile. During the first flight—a campaign of

The badge from the last mission of the *Endeavour* shuttle, which hosted an experiment from our laboratory.

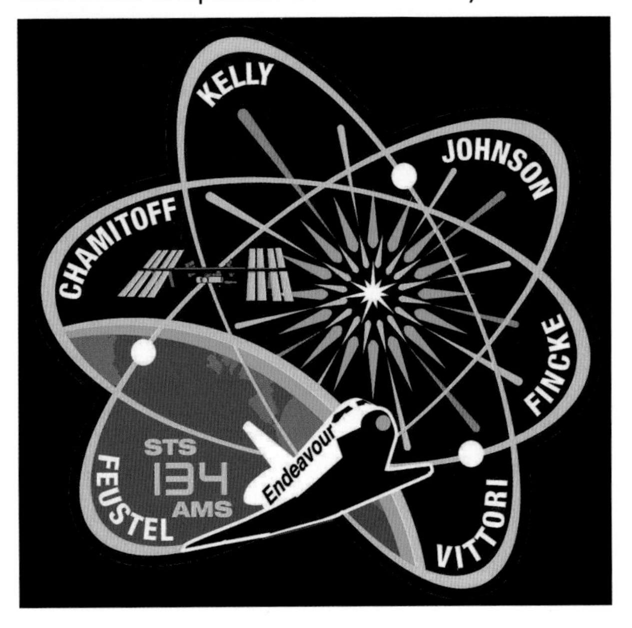

parabolic flights requires three, on successive days—I had to assess the functioning of the instruments and, if everything worked properly—something I was very dubious about—I could carry out some experiments. The plane took off and went straight over the Atlantic, where it began the sequence of thirty parabolas. During each one we had about twenty seconds of weightlessness.

Each parabola begins with a phase of ascension in which the plane rises for about thirty seconds at an angle of about forty-five degrees and at a very high velocity, subjecting the passengers to an acceleration of almost 2 g's (as if you weighed twice as much as normal). At the culmination of the acceleration, the pilot stops stoking the engines and begins the so-called ballistic flight, during which the aircraft becomes a bullet fired into the atmosphere and weightlessness begins. The transition from double gravity to zero gravity is immediate. Your body detaches from the ground and begins to float in the air. Above and below lose any meaning, and every movement becomes unnatural. Some liken the absence of gravity to floating in water, others to falling off a cliff. It is a feeling that cannot be described, because it is unlike anything that a living being has ever experienced during its lifetime. It is such a novel sensation that during the night after the flight, those who have experienced weightlessness for the first time often dream that they are experiencing it again. That is the brain trying to sort the abnormal sensation into the context of past experiences. The fact is that it is very pleasant. To weigh nothing, to float in the air, walk on the ceiling of the plane, twist endlessly. Like other firsts in your life, you never forget your first parabola.

Then the pilot pumps the engines, and from the spirit you were, you become matter once more. To my surprise the experiments went smoothly and, from the first parabola, even though I used my time almost exclusively to experience weightlessness, the computers began recording some significant action potentials (electrical signals similar to those flowing between the neurons of our brain) in the plant, occurring

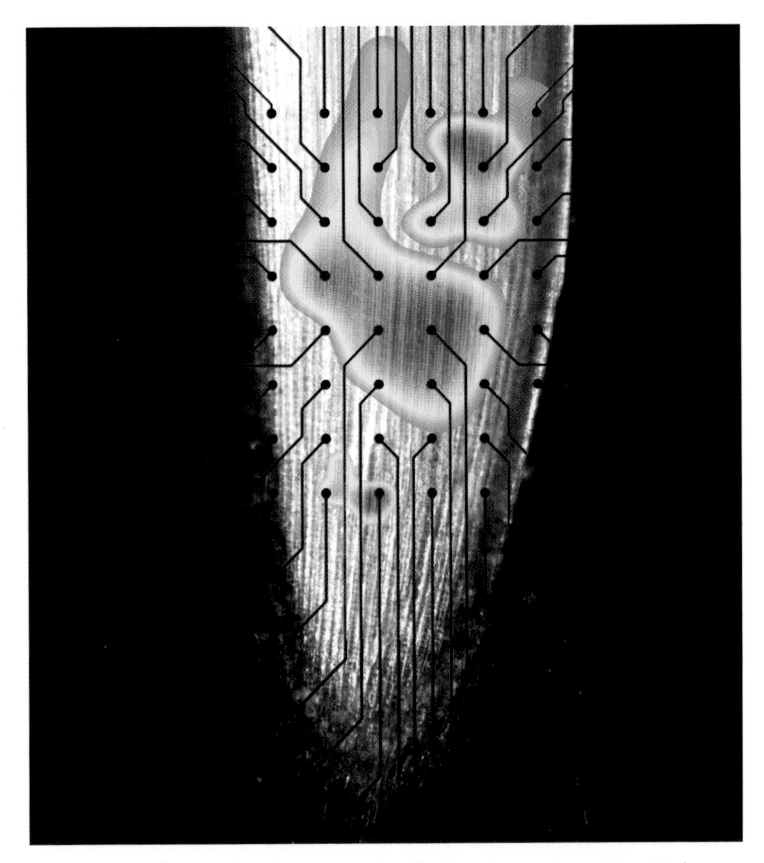

A root apex of corn placed on a matrix of microelectrodes, with which to measure action potentials generated spontaneously by the cells.

in the area that we had expected and from there spreading to nearby regions of the root apex. I did not know it at that point, but we were measuring what would turn out to be the fastest signals produced by a plant in response to weightlessness. Just one and a half seconds after the start of microgravity, action potentials were produced in the root and then moved to adjacent regions. It was an exceptional result: until then the fastest signal ever recorded had taken place about ten minutes after the start of microgravity.

Once again the plant demonstrated a sense capability that was far greater than any we had imagined. To know that the root was responding so quickly to changes in gravity opened new perspectives. We had perhaps found the first event that, through a multitude of physiological adaptations that followed, allowed the plant to adjust to gravitational conditions other than those found on Earth. It was a promising first step that, together with the discoveries of many other scientists engaged in the study of space biology, will one day soon enable us to understand how plants, the masters of resistance and adaptation, also adapt to the absence of gravity.

In awe, I watched the signals repeat with regularity during every subsequent parabola. The data collected demonstrated plants' extreme

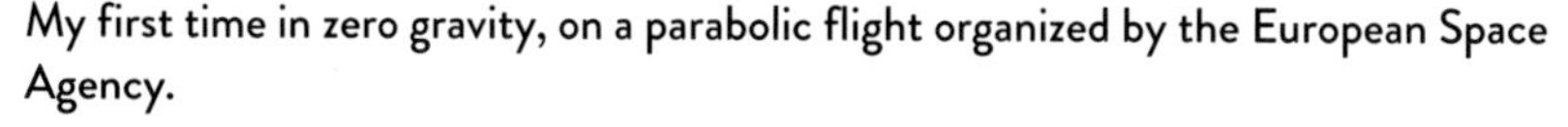

My first time in zero gravity, on a parabolic flight organized by the European Space Agency.

reactivity and helped convince the scientific community and space agencies that as experimental subjects, they were more than suited to the study conditions of parabolic flights.

In later years, I participated in other similar campaigns, some with even better results. But it is my first parabolic flight that will remain foremost in my memory. It was a moment of pure happiness. On the same day and during the same flight, I had floated weightless for the first time and recorded the fastest known signal emitted from a plant in response to a decrease in gravity.

That is the kind of moment that every scientist dreams of experiencing, an enduring reminder of how wonderful the life of a scientist can be.

References and Sources

PREFACE

RGB Kew. *State of the World's Plants, 2016.* https://stateoftheworldsplants.com/report/sotwp_2016.pdf.

Risen, C. C. "The World's Most Advanced Building Material Is . . . Wood. And It's Going to Remake the Skyline." *Popular Science*, February 6, 2014.

CHAPTER 1

Chakrabortee, S., C. Kayatekin, G. A. Newby, et al. "Luminidependens (LD) Is an *Arabidopsis* Protein with Prion Behavior." *Proceedings of the National Academy of Sciences of the United States of America* 113, no. 21 (2016): 6065–70.

Gagliano, M., M. Renton, M. Depczynski, and S. Mancuso. "Experience Teaches Plants to Learn Faster and Forget Slower in Environments Where It Matters." *Oecologia* 175, no. 1 (May 2014): 63–72.

Hawkes, E. J., S. P. Hennelly, I. V. Novikova, et al. "COOLAIR Antisense RNAs from Evolutionarily Conserved Elaborate Secondary Structures." *Cell Reports* 16, no. 12 (September 20, 2016): 3087–96.

Lamarck, J. B., and A. P. de Candolle. "*Flore française, ou, Descriptions succinctes de toutes les plantes qui croissent naturellement en France* [French Flora or Short Summaries of All the Plants That Naturally Grow in France]." Paris: Desray, 1815.

CHAPTER 2

Brenner, E. D., R. Stahlberg, S. Mancuso, et al. "Plant Neurobiology: An Integrated View of Plant Signaling." *Trends in Plant Science* 11, no. 8 (August 11, 2006): 413–19.

Darwin, F. "The Address of the President of the British Association for the Advancement of Science." *Science—New Series* 716, no. 28 (September 18, 1908): 353–62.

Dawson, C., J. F. V. Vincent, and A.-M. Rocca. "How Pine Cones Open." *Nature* 390, no. 1661 (December 18, 1997): 668.

Greacen, E. L., and J. S. Oh. "Physics of Root Growth." *Nature New Biology* 235, no. 53 (January 5, 1972): 24–25.

Ma, M., L. Guo, D. G. Anderson, and R. Langer. "Bio-inspired Polymer Composite Actuator and Generator Driven by Water Gradients." *Science* 339, no. 6116 (January 11, 2013): 186–89.

Mancuso, S., B. Mazzolai, D. Comparini, et al. *Subsurface Investigation and Interaction by Self-burying Bio-inspired Probes: Self-burial Strategy and Performance in* Erodium cicutarium*—SeeDriller*. Final report, December 2, 2014. https://www.esa.int/gsp/ACT/doc/ARI/ARI%20Study%20Report/ACT-RPT-BIO-ARI-12-6401-selfburying.pdf.

Mancuso, S., and B. Mazzolai. "Il plantoide: Un possibile prezioso robot ispirato al mondo vegetale [Plantoids: A Potentially Precious Robot Inspired by the Plant World]." In *Atti dei Georgofili 2006*. Firenze: Accademia dei Georgofili, 2007: 223–34.

Mazzolai, B., C. Laschi, P. Dario, et al. "The Plant as a Biomechatronic System." *Plant Signaling & Behavior* 5, no. 2 (February 2010): 90–93.

McClung, C. R. "Plant Circadian Rhythms." *The Plant Cell* 18, no. 4 (April 2006): 792–803.

Murawski, D. "Genetic Variation Within Tropical Tree Crowns." In *Biologie d'une canopée de forêt équatoriale* [Biology of an Equatorial Forest Canopy]. Vol. 3: *Rapport de la mission d'exploration scientifique de la canopée de Guyane, octobre–décembre 1996* [Report on the Mission of Scientific Exploration of the Guyana Canopy, October–December 1996]. Ed. F. Hallé. Paris: Pro-Natura International, 1998.

CHAPTER 3

Baluška, F., and S. Mancuso. "Vision in Plants via Plant-Specific Ocelli?" *Trends in Plant Science* 21, no. 9 (September 2016): 727–30.

Benbrook, C. M. "Trends in Glyphosate Herbicide Use in the United States and Globally." *Environmental Sciences Europe* 28, no. 3 (February 2, 2016).

Darwin, F. "Lectures on the Physiology of Movement in Plants. V. The Sense-organs for Gravity and Light." *New Phytologist* 6, no. 3/4 (1907): 69–76.

Gavelis, G. S., S. Hayakawa, R. A. White 3rd, et al. "Eye-like Ocelloids Are Built from Different Endosymbiotically Acquired Components." *Nature* 523, no. 7559 (July 9, 2015): 204–07.

Haberlandt, G. *Die Lichtsinnesorgane der Laubblätter*. Leipzig: Engelmann Verlag, 1905.

Hamilton, W. D., and S. P. Brown. "Autumn Tree Colors as a Handicap Signal." *Proceedings of the Royal Society B* 268, no. 1475 (July 22, 2001): 1489–93.

Hayakawa, S., Y. Takaku, J. S. Hwang, et al. "Function and Evolutionary Origin of Unicellular Camera-type Eye Structure." *PLoS One* 10, no. 3 (March 3, 2015).

Mancuso, S. *Uomini che amano le piante: Storie di scienziati del mondo vegetale* [Men Who Love Plants]. Florence and Milan: Giunti, 2014.

Mancuso, S., and A. Viola. *Brilliant Green: The Surprising History and Science of Plant Intelligence*. Washington, DC: Island Press, 2013.

Mancuso, S., and A. Viola. *Verde brillante: Sensibilità e intelligenza del mondo vegetale*. Florence and Milan: Giunti, 2013.

Mancuso, S., and F. Baluška. "Plant Ocelli for Visually Guided Plant Behavior." *Trends in Plant Science* 22, no. 1 (January 2017): 5–6.

Schuergers, N., T. Lenn, R. Kampmann, et al. "Cyanobacteria Use Micro-optics to Sense Light Direction." *eLife* 5, February 9, 2016.

Vavilov, N. I. *Origin and Geography of Cultivated Plants*. Cambridge, UK: Cambridge University Press, 1992.

Wager, H. "The Perception of Light in Plants." *Annals of Botany* 23, no. 3 (July 1, 1909): 459–90.

CHAPTER 4

Arrow, K. J., R. Forsythe, M. Gorham, et al. "Economics: The Promise of Prediction Markets." *Science* 320, no. 5878 (May 16, 2008): 877–78.

Baluška, F., S. Lev-Yadun, and S. Mancuso. "Swarm Intelligence in Plant Roots." *Trends in Ecology and Evolution* 25, no. 12 (December 2010): 682–83.

Bonabeau, E., M. Dorigo, and G. Theraulaz. *Swarm Intelligence: From Natural to Artificial Systems*. New York: Oxford University Press, 1999.

Borges, J. L. "El idioma analítico de John Wilkins [The Analytical Language of John Wilkins]." In *Otras inquisiciones (1937–1952)*. Buenos Aires: Sur, 1952.

Ciszak, M., D. Comparini, B. Mazzolai, et al. "Swarming Behavior in Plant Roots." *PLoS One* 7, no. 1 (2012).

Clément, R. J. G., S. Krause, N. von Engelhardt, et al. "Collective Cognition in Humans: Groups Outperform Their Best Members in a Sentence Reconstruction Task." *PLoS One* 8, no. 10 (October 17, 2013).

Conradt, L., and T. J. Roper. "Group Decision-making in Animals." *Nature* 421, no. 6919 (January 9, 2003): 155–58.

Couzin, I. D. "Collective Cognition in Animal Groups." *Trends in Cognitive Sciences* 13, no. 1 (January 2009): 36–43.

Eshel, A., and T. Beeckman, eds. *Plant Roots: The Hidden Half*, 4th ed. Boca Raton, FL: CRC Press, 2013.

Gigerenzer, G. *Gut Feelings: The Intelligence of the Unconscious*. New York: Viking, 2007.

Hallé, F. *Éloge de la plante: Pour une nouvelle biologie*. Paris: Seuil, 1999.

Kerr, N. L., and R. S. Tindale. "Group Performance and Decision Making." *Annual Review of Psychology* 55 (February 4, 2004): 623–55.

Klein, N., and N. Epley. "Group Discussion Improves Lie Detection." *Proceedings of the National Academy of Sciences of the United States of America* 112, no. 24 (June 16, 2015): 7460–65.

Krause, J., G. D. Ruxton, and S. Krause. "Swarm Intelligence in Animals and Humans." *Trends in Ecology & Evolution* 25, no. 1 (January 2010): 28–34.

Kurvers, R. H., J. Krause, G. Argenziano, et al. "Detection Accuracy of Collective Intelligence Assessments for Skin Cancer Diagnosis." *JAMA Dermatology* 151, no. 12 (December 1, 2015): 1346–53.

Kurvers, R. H. J. M., S. M. Herzog, R. Hertwig, et al. "Boosting Medical Diagnostics by Pooling Independent Judgments." *Proceedings of the National*

Academy of Sciences of the United States of America 113, no. 31 (August 16, 2016): 8777–82.

Lakhani, K. R., D. A. Garvin, and E. Lonstein. "TopCoder (A): Developing Software Through Crowdsourcing." Harvard Business School Case 610-032, January 2010.

Mellers, B., L. Ungar, J. Baron, et al. "Psychological Strategies for Winning a Geopolitical Forecasting Tournament." *Psychological Science* 25, no. 5 (May 1, 2014): 1106–15.

Pfeiffer, T., and J. Almenberg. "Prediction Markets and Their Potential Role in Biomedical Research: A Review." *Biosystems* 102, no. 2/3 (November–December 2010): 71–76.

Plato. *Protagoras*. Trans. Benjamin Jowett. http://classics.mit.edu/Plato/protagoras.html.

Surowiecki, J. *The Wisdom of Crowds: Why the Many Are Smarter than the Few and How Collective Wisdom Shapes Business, Economies, Societies and Nations*. New York: Doubleday, 2004.

Wolf, M., J. Krause, P. A. Carney, et al. "Collective Intelligence Meets Medical Decision-Making: The Collective Outperforms the Best Radiologist." *PLoS One* 10, no. 8 (August 12, 2015).

Wolf, M., R. H. J. M. Kurvers, A. J. W. Ward, et al. "Accurate Decisions in an Uncertain World: Collective Cognition Increases True Positives While Decreasing False Positives." *Proceedings of the Royal Society B* 280, no. 1756 (2013): 2012–77.

Woolley, A. W., C. F. Chabris, A. Pentland, et al. "Evidence for a Collective Intelligence Factor in the Performance of Human Groups." *Science* 330, no. 6004 (October 29, 2010): 686–88.

CHAPTER 5

Boecker, H., T. Sprenger, M. E. Spilker, et al. "The Runner's High: Opioidergic Mechanisms in the Human Brain." *Cerebral Cortex* 18, no. 11 (2008): 2523–31.

Byrnes, N. K., and J. E. Hayes. "Personality Factors Predict Spicy Food Liking and Intake." *Food Quality and Preference* 28, no. 1 (April 1, 2013): 213–21.

Delpino, F. "Rapporti tra insetti e nettari extranuziali nelle piante [The

Relationship Between Insects and Extrafloral Nectaries]." *Bollettino della Società Entomologica Italiana* [Italian Entomological Society Bulletin] 6 (1874): 234–39.

Nepi, M. "Beyond Nectar Sweetness: The Hidden Ecological Role of Non-protein Amino Acids in Nectar." *Journal of Ecology* 102, no. 1 (January 2014): 108–15.

Nicolson, S. W., and R. W. Thornburg. "Nectar Chemistry." In *Nectaries and Nectar*, ed. S. W. Nicolson, M. Nepi, and E. Pacini. Dordrecht: Springer, 2007: 215–64.

Rico-Gray, V., and P. S. Oliveira. *The Ecology and Evolution of Ant-Plant Interactions*. Chicago: University of Chicago Press, 2007.

Schoonhoven, L. M., J. J. A. van Loon, and M. Dicke. *Insect-Plant Biology*, 2nd ed. Oxford, UK: Oxford University Press, 2005.

Scoville, W. L. "Note on Capsicums." *The Journal of the American Pharmaceutical Association* 1, no. 5 (May 1912): 453–54.

CHAPTER 6

Hooker, J. D. *Life and Letters of Sir Joseph Dalton Hooker*, vol. 2. L. Huxley, ed. London: John Murray, 1918: 25.

Hooker, J. D. "On *Welwitschia*, a New Genus of Gnetaceae." *The Transactions of the Linnean Society of London* 24, no. 1 (January 1863): 1–48.

Ju, J., H. Bai, Y. Zheng, et al. "A Multi-structural and Multi-functional Integrated Fog Collection System in Cactus." *Nature Communications* 3, no. 1247 (2012).

Leonardo da Vinci. *Trattato della pittura* [Treatise on Painting]. Parte VI: *Degli alberi e delle verdure* [On Trees and Greenery]. N. 833: *Della scorza degli alberi* [On the Bark of Trees]. Rome: Newton Compton, 2015.

Nørgaard, T., and M. Dacke. "Fog-Basking Behaviour and Water Collection Efficiency in Namib Desert Darkling Beetles." *Frontiers in Zoology* 7, no. 23 (July 16, 2010).

Zheng, Y., H. Bai, Z. Huang, et al. "Directional Water Collection on Wetted Spider Silk." *Nature* 463, no. 7281 (February 4, 2010): 640–43.

CHAPTER 7

Grassini, P., K. M. Eskridge, and K. G. Cassman. "Distinguishing Between Yield Advances and Yield Plateaus in Historical Crop Production Trends." *Nature Communications* 4, no. 2918 (2013).

Kelley, C. P., S. Mohtadi, M. A. Cane, et al. "Climate Change in the Fertile Crescent and Implications of the Recent Syrian Drought." *Proceedings of the National Academy of Sciences of the United States of America* 112, no. 11 (March 17, 2015): 3241–46.

Lesk, C., P. Rowhani, and N. Ramankutty. "Influence of Extreme Weather Disasters on Global Crop Production." *Nature* 529, no. 7584 (January 7, 2016): 84–87.

Qadir, M., A. Tubeileh, J. Akhtar, et al. "Productivity Enhancement of Salt-Affected Environments Through Crop Diversification." *Land Degradation & Development* 19, no. 4 (June 20, 2008): 429–53.

Riadh, K., M. Wieded, K. Hans-Werner, and A. Chedly. "Responses of Halophytes to Environmental Stresses with Special Emphasis to Salinity." *Advances in Botanical Research* 53 (December 2010): 117–45.

Ricardo, D. *On the Principles of Political Economy and Taxation*. New York: Cosimo, 2006: 46–47.

Ruan, C.-J., J. A. T. da Silva, S. Mopper, et al. "Halophyte Improvement for a Salinized World." *Critical Reviews in Plant Sciences* 29, no. 6 (2010): 329–59.

Transnational Institute for European Coordination Via Campesina and Hands Off the Land Network. *Land Concentration, Land Grabbing and People's Struggles in Europe*. 2013. https://www.tni.org/files/download/land_in_europe-jun2013.pdf.

Wallace, D. F. *This Is Water*. Boston: Little, Brown and Company, 2009.

World Economic Forum. *The Global Risks Report 2016*. http://www3.weforum.org/docs/GRR/WEF_GRR16.pdf.

CHAPTER 8

Baluška, F., S. Mancuso, and D. Volkmann, eds. *Communication in Plants: Neuronal Aspects of Plant Life*. Berlin: Springer, 2006.

Braun, A. C. H. "The Vegetable Individual, in Its Relation to Species." *The American Journal of Science and Arts* 19 (1855): 297–318.

Clark, L. J., W. R. Whalley, and P. B. Barraclough. "How Do Roots Penetrate Strong Soil?" *Plant and Soil* 255, no. 1 (August 2003): 93–104.

Darwin, C. *Journal of Researches into the Geology and Natural History of the Various Countries Visited by H. M. S.* Beagle, *Under the Command of Captain Fitzroy, R.N., from 1832 to 1836*. London: Henry Colburn, 1839.

Darwin, C. *The Narrative of the Voyages of H. M. Ships* Adventure *and* Beagle. Vol. 3, *Journal and Remarks, 1832–1836* (*The Voyage of the* Beagle). London: Henry Colburn, 1839.

Darwin, E. *Phytologia, or, the Philosophy of Agriculture and Gardening*. London: Johnson, 1800: 2.

Fabre, J. H. *La plante: Leçons à mon fils sur la botanique* [Plants: Lessons on Botany for My Son]. Paris: Librairie Charles Delagrave, 1876.

Goethe, J. W. von. *Versuch die Metamorphose der Pflanzen zu erklären* [Metamorphosis of Plants]. Gotha: C. W. Ettinger, 1790.

CHAPTER 9

Armstrong, N., M. Collins, and E. E. Aldrin, Jr. *First on the Moon: A Voyage with Neil Armstrong, Michael Collins, Edwin E. Aldrin Jr.* Boston: Little, Brown and Company, 1970.

Asimov, I. *Pebble in the Sky*. New York: Doubleday, 1950.

Baluška, F., S. Mancuso, D. Volkmann, and P. W. Barlow. "Root Apex Transition Zone: A Signalling-Response Nexus in the Root." *Trends in Plant Science* 15, no. 7 (July 2010): 402–08.

Barlow, P. W. "Gravity Perception in Plants: A Multiplicity of Systems Derived by Evolution?" *Plant, Cell & Environment* 18, no. 9 (September 1995): 951–62.

Hu, E., S. I. Bartsev, and H. Liu. "Conceptual Design of a Bioregenerative Life Support System Containing Crops and Silkworms." *Advances in Space Research* 45, no. 7 (April 1, 2010): 929–39.

Masi, E., M. Ciszak, D. Comparini, et al. "The Electrical Network of Maize Root Apex Is Gravity Dependent." *Scientific Reports* 5, article no. 7730 (2015).

Pandolfi, C., E. Masi, B. Voigt, et al. "Gravity Affects the Closure of the Traps in *Dionaea muscipula*." *BioMed Research International*, 2014. https://www.hindawi.com/journals/bmri/2014/964203/.

Photo Credits

Unless otherwise indicated, the images come from the Giunti Archive. The publisher will settle any reproduction rights for those images for which it has not been able to find the source.

pp. xiii, 12, 14, 18–19, 51r: Courtesy of Stefano Mancuso
pp. 2–3 © Peter Owen/EyeEm/Getty Images
p. 7 © TK
p. 9 © Shutterstock/Bankolo5
p. 22 © SSPL/National Media Museum/Getty Images
p. 24 © SSPL/Florilegius/Getty Images
p. 46 © Flickr/Wikimedia Commons
p. 48 left © blickwinkel/Alamy Stock Photo/IPA
p. 48 right © Shutterstock/vaivirga
p. 51 left © Paul Zahl/National Geographic/Getty Images
p. 57 © Shutterstock/ChWeiss
p. 58 © UIG/Getty Images
p. 61 © Pal Hermansen/NPL/Contrasto
p. 64 © Bill Barksdale/Agefotostock
p. 79 © Shutterstock/Hristo Rusev
p. 81 © Stuart Wilson/Science Source/Getty Images
p. 88 © Martin Ruegner/IFA-Bilderteam/Getty Images
p. 91 © The Opte Project/Wikimedia Commons
p. 103 © Morley Read/Getty Images

p. 106 © Visuals Unlimited/NPL/Contrasto
p. 112 © Shutterstock/Lenscap Photography
p. 128 © Hulton Deutsch/Getty Images
p. 129 © George Rinhart/Getty Images
p. 137 © UIG/Getty Images
p. 140 © Juan Carlos Munoz/NPL/Contrasto
pp. 148–149 © Shutterstock/Peter Wey
p. 152 © Andrii Shevchuk/Alamy Stock Photo/IPA
p. 155 © Shutterstock/Atonen Gala
p. 156 © Shutterstock/Darren J. Bradley
p. 157 © Peter Chadwick/SPL/Contrasto
p. 173 top © De Agostini/Getty Images
p. 173 bottom © Lebrecht Music & Arts/Contrasto
p. 178 © Labrina/Creative Commons

Index

B

C

E

F

G

P

T

U

V

W